Modeling and Optimization in Science and Technologies

Volume 13

The book series *Modeling and Optimization in Science and Technologies* (*MOST*) publishes basic principles as well as novel theories and methods in the fast-evolving field of modeling and optimization. Topics of interest include, but are not limited to: methods for analysis, design and control of complex systems, networks and machines; methods for analysis, visualization and management of large data sets; use of supercomputers for modeling complex systems; digital signal processing; molecular modeling; and tools and software solutions for different scientific and technological purposes. Special emphasis is given to publications discussing novel theories and practical solutions that, by overcoming the limitations of traditional methods, may successfully address modern scientific challenges, thus promoting scientific and technological progress. The series publishes monographs, contributed volumes and conference proceedings, as well as advanced textbooks. The main targets of the series are graduate students, researchers and professionals working at the forefront of their fields.

More information about this series at http://www.springer.com/series/10577

Héctor Benítez-Pérez
Jorge L. Ortega-Arjona
Paul E. Méndez-Monroy
Ernesto Rubio-Acosta
Oscar A. Esquivel-Flores

Control Strategies and Co-Design of Networked Control Systems

Considering Time Delay Effects

Héctor Benítez-Pérez
Departamento de Ingeniería de Sistemas Computacionales y Automatización
IIMAS—National Autonomous University of Mexico
Mexico City, Mexico

Ernesto Rubio-Acosta
Departamento de Ingeniería de Sistemas Computacionales y Automatización
IIMAS—National Autonomous University of Mexico
Mexico City, Mexico

Jorge L. Ortega-Arjona
Departamento de Matemáticas Facultad de Ciencias
National Autonomous University of Mexico
Mexico City, Mexico

Oscar A. Esquivel-Flores
Departamento de Ingeniería de Sistemas Computacionales y Automatización
IIMAS—National Autonomous University of Mexico
Mexico City, Mexico

Paul E. Méndez-Monroy
Departamento de Ciencias de la Computación
IIMAS Mérida—National Autonomous University of Mexico
Mérida, Yucatán, Mexico

ISSN 2196-7326 ISSN 2196-7334 (electronic)
Modeling and Optimization in Science and Technologies
ISBN 978-3-030-07290-2 ISBN 978-3-319-97044-8 (eBook)
https://doi.org/10.1007/978-3-319-97044-8

Softcover re-print of the Hardcover 1st edition 2019

This Springer imprint is published by the registered company Springer Nature Switzerland AG
The registered company address is: Gewerbestrasse 11, 6330 Cham, Switzerland

To Rita

—Héctor Benítez-Pérez

To Lucía and Jorgito

—Jorge L. Ortega-Arjona

To Lia, Maylen and Mateo

—Paul E. Méndez-Monroy

To Lizbeth

—Ernesto Rubio-Acosta

To my dear Sofia G.

—Oscar A. Esquivel-Flores

To all our families ...

Preface

Network systems are intended to communicate nodes which simultaneously execute. Today, these nodes are autonomous machines, on which a single program may be executed (called a "distributed program"). Each node works on a part of the problem. Nevertheless, nodes proceed together through time, exchanging data, and sharing resources towards a common objective. Nowadays, network computing attempts to become an important mainstream for computing. In recent years, the most powerful computer system is a networked one. A reason for this is that once computer manufacturers have built a powerful computer, two of them can be networked to obtain a more powerful system. A distributed system can be defined, in general terms, as the specification of a set of processes executing simultaneously and communicating among themselves in order to achieve a common objective.

This book presents the networked control system (NCS) as a particular kind of a real-time distributed system (RTDS), composed of a set of nodes, interconnected by a network, able to develop a complete control process. Important parts of such a control process are sensor and actuator activities, which rely on a real-time operating system and a real-time communication network. NCSs have received increasing attention over the last few years, due to several advantages, such as an easy maintenance, a high reliability, less wiring, and low cost. Besides, the use of control network architectures, when properly applied, may improve efficiency, flexibility, and reliability, reducing installation, reconfiguration, maintenance in time and costs. Nevertheless, notice that the use of a single communication network for all process control signals (e.g. the feedback control loop) affects the performance of the whole NCS. This is because the network generates some imperfections, such as time delays, packet loss. Furthermore, conventional standard results in analysis and design of digital control theory cannot be used, since many of its assumptions are not directly applicable to NCS, specifically considering time delays and packet loss. The use of common-bus network architecture introduces different forms of uncertainties between sensors, actuators, and controllers. Several approaches have been developed to tackle this problem. In such approaches, a group of nodes make use of the same network, producing a traffic load. If there is no coordination among the nodes, data transmissions may simultaneously occur, generating time delays, retransmissions, or worst packet loss. This results in tasks which are unable to fulfil their deadlines. In order to achieve overall objectives of all tasks performed in an NCS, it is necessary for all nodes to properly exchange their own information through the network. Therefore, the mechanism to exchange information plays an important role in the stability and performance of the NCS implemented over a communication network.

In an NCS, the main problems that degrade its performance are network-induced delays and packet loss. Time delays can be constant, time varying, or even random; those depend on the scheduler, network type, architecture, operating systems, etc. When an NCS uses a reliable network, it is possible to compensate for time delays, lower or larger than the sampling period, and obtain bounds. However, when an NCS uses an unreliable network, it is more difficult to compensate for time delays and packet loss, and these reduce the performance of the NCS.

Reconfiguration is a transition that modifies the structure of a system, so it changes its state representation [1]. In this book, it is used as a feasible approach for fault isolation, and also, it is used to deal with time delay modification. In control systems, several modelling strategies for managing time delays within control laws have been studied by different research groups. A classical definition of an NCS is that it is a dynamic system that makes use of a network to be controlled through it. NCSs have been studied for the last 20 years, using several assumptions regarding time delays modelling, or uncertainties accomplishment. In this book, some network imperfections are modelled, in order to be coupled with a control law design. Several strategies are reviewed, such as stochastic approximation, communication frequency modelling, or dynamic scheduling procedure. All of them trigger certain threshold onto dynamic systems models. However, the modelling procedure for NCSs is challenging, since there are several nonlinear situations, like local saturations, uncertain time delays, dead zones, or local situations, with which it is necessary to deal with. A strong configuration for modelling and control has been used through this book, like fuzzy control design. A fuzzy Takagi–Sugeno control is designed following bounded time delays, with a guarantee of certain known behaviour, which is the result of real-time behaviour or "normal" operation. Other static strategies, like digital control, are unable to produce a feasible result since time delays are variable. Moreover, time delay variation cannot be modelled as a result of uncertainties, since the system structure is modified upon certain variation.

The approach proposed for this book has the following major features:

1. NCSs are modelled through a powerful technique, like the fuzzy Takagi–Sugeno approach.
2. There is a proper approximation of system design and time delay representations, allowing to perform a suitable and stable modelling of several local problems that, otherwise, may terminate into catastrophic situations.
3. Local time delays are modelled by a fundamental understanding of communication interactions among nodes, as well as computing processing through scheduling restrictions.

4. Several examples are presented to show congruent results from several kinds of implementations, based on a LVDT system, a magnetic levitator system, and a helicopter representation.

The structure of the book is briefly described as follows:

- **Chapter** 1 **Introduction to Networked Control Systems** presents a brief description of networked control systems. A formal review of several methodologies for NCS from control, scheduling, and codesign is presented. Also, it presents a description of sources of uncertainties generated by network, their identification, modelling, and bounding. Two uncertainties are studied in this book, the time delays smaller or larger than a sampling period and lost packet. Finally, a detailed review of this book is given.
- **Chapter** 2 **Modelling of Networked Control Systems** describes some statistical representations of different sources of time delays and how they are integrated into state-space models. Several models are presented, showing how time delays are bounded. Two fuzzy models represent the system dynamics with several time delays and lost packets. Also, a frequency model represents the transmission dynamics into a network. Finally, a neuro-fuzzy model is used to identify the system dynamics.
- **Chapter** 3 **Distributed Systems Modelling** reviews some representations of time delays in function of some features as the operative system, network, scheduling. These include the scalability, concurrency, feasibility, and extensibility. At the end, an introduction to a example of dynamic model is given, focusing on real-time representation with a relationship matrix of transmission frequencies through true-time simulation.
- **Chapter** 4 **Design of Networked Control Systems** presents the control strategies and codesign based on the NCS models. Three methodologies are proposed focusing on vanishing the uncertainties of time delays larger than a sampling period and lost packets. First, a LQR control is designed for the adaptive fuzzy model with the known and bounded time delay. A sampling frequency control modifies the sampling frequencies into a region according the quality of service. Finally, a codesign methodology is presented where the QoS and QoC are trade-off with two fuzzy systems considering the time delays larger than the sampling period and lost packets.
- **Chapter** 5 **Control Design Considering Mobile Computing** presents an extension of distributed systems considering mobile computing, a review of some scheduling algorithms is presented to get an approximation to bounded time in all stages of communication and compute, and also it has a review of the most important algorithms in terms of consensus and routing. Finally, a computer network design is presented from a selection of real-time features, scheduler, task handlers, priorities, precedencies, and consensus.
- **Chapter** 6 **Applications** shows several applications of the proposed design strategies for NCS. Two case studies are described—a SISO MAGLEV system and MIMO 2-DOF helicopter system—both are used in different NCS configurations in the chapter to prove the effectiveness of the proposed methods. The

three methodologies presented in the previous chapter are designed for the case studies. All of them show the versatility in the design with multiple uncertainties and a fully distributed configuration with an excellent system performance. The NCS codesign shows that the combined application of strategies improves the performance of the complete system.

- Finally, **Chap**. 7 **Conclusions** presents the conclusions of this book emphasizing the multiple strategies. Their advantages about the NCS configuration, task handlers, control, and scheduling algorithms. Further, this highlights the qualities of fuzzy systems for control and scheduling of NCS.

Mexico City, Mexico — Héctor Benítez-Pérez
Mexico City, Mexico — Jorge L. Ortega-Arjona
Mexico City, Mexico — Ernesto Rubio-Acosta
Mérida, Mexico — Paul E. Méndez-Monroy
Mexico City, Mexico — Oscar A. Esquivel-Flores
June 2018

Reference

1. Benitez-Perez, H., Garcia-Zavala, A., Garcia-Nocetti, F.: A proposal for online reconfiguration based upon a modification of planning scheduling and fuzzy logic control law response. ISSADS, Lecture Notes Computer Science, pp. 141–152. Springer (2005)

Acknowledgements

The authors wish to express their gratitude to the Consejo Nacional de Ciencia y Tecnologia (CONACYT) and the Universidad Nacional Autonoma de Mexico (UNAM), as the main sponsors of the research projects through the CONACYT 176556, PAPIIT IA104218, and PAPIIT 1N100813 projects, respectively, whose outcome through years is the base of these pages. Furthermore, acknowledgements are due to all the contributors of this book, particularly Angel Garcia-Zavala, Magali Arellano-Vazquez, Jose Angel Hermosillo-Gomez, Jared Rojas-Vargas, Adrian Duran-Chavesti, Jose Alberto Oriol Castillo, and Andres Alvarez-Cid, without whose effort and enthusiasm all this work would not be.

Contents

About the Authors

Héctor Benítez-Pérez graduated with an honourable mention in his B.Eng. degree in Electrical Engineering from the Universidad Nacional Autónoma de México (UNAM) (1989–1993) and received a Ph.D. degree in Automatic Control and Systems Engineering from the University of Sheffield (1995–1999). Currently, he is Tenured Researcher at the Instituto de Investigaciones en Matemáticas Aplicadas y en Sistemas (IIMAS), Director in IIMAS since 2012, Member of the National System of Researchers, Level II, and Member of the Mexican Academy of Sciences and Engineering. His research interests include networked control and distributed systems.

Jorge L. Ortega-Arjona is Full-time Assistant Professor in the Department of Mathematics, Faculty of Science, Universidad Nacional Autónoma de México (UNAM). He received a B.Eng. degree in Electronic Engineering, an M.Sc. degree in Computer Science from UNAM, and a Ph.D. degree in Computer Science from the University College London. His current research interests include parallel software design, parallel processing, object-oriented programming software patterns, and software design and architecture.

Paul E. Méndez-Monroy received a B.Eng. degree in Communications and Electronic Engineering from the Instituto Politécnico Nacional and the MI degree and Ph.D. degree in Electrical Engineering from the Universidad Nacional Autónoma de México (UNAM). Currently, he is a Full-time Associate Researcher in the Instituto de Investigaciones en Matemáticas Aplicadas y en Sistemas (IIMAS) Mérida campus at UNAM and Member of the National System of Researchers, Level I. His research interests include machine learning, robotics, distributed and parallel computing, and related areas.

Ernesto Rubio-Acosta is Full-time Assistant Researcher in the Instituto de Investigaciones en Matemáticas Aplicadas y en Sistemas (IIMAS) at the Universidad Nacional Autónoma de México (UNAM). He received a B.Eng. degree in Mechanical and Electrical Engineering, an M.Sc. degree in Computer Science, and a Ph.D. degree in Earth Science, all of them from UNAM. Currently, his research interests include the mathematical modelling of underground flow systems and their numerical solution using high-performance parallel computing.

Oscar A. Esquivel-Flores received the B.Sc. degree in Applied Mathematics and Computing from the Universidad Nacional Autónoma de México (UNAM), the M.Sc. degree in Computer Sciences from the Universidad Autónoma Metropolitana (UAM), and a Ph.D. degree in Computer Engineering from UNAM. He was Postdoc Researcher on parallel and high-performance computing at the Barcelona Supercomputing Center. Currently, he is Postdoc Researcher on data scientist at the Tecnológico de Monterrey and Member of the National System of Researchers level candidate.

Acronyms

A/D	Analog/digital
AC	Alternating current
ACK	Acknowledgement
AI	Artificial intelligence
AODV	Ad hoc on-demand distance vector
API	Application programming interface
BB	Blackboard
BDI	Beliefs–Desires–Intentions
BEB	Binary exponential backtracking
CAN	Controller area network
CP-TOD	Constant penalty try to discard
CPU	Central processing unit
CRC	Cyclic redundancy check
CSMA/AMP	Carrier sense multiple access/arbitration by message priority
CSMA/BA	Carrier sense multiple access/bitwise arbitration
CSMA/CA	Carrier sense multiple access/collision avoidance
CSMA/CD	Carrier sense multiple access/collision detection
CSMA/NBA	Carrier sense multiple access/non-destructive bitwise arbitration
CT	Coordinated task
D/A	Digital/analog
DAI	Distributed artificial intelligence
DAQ	Data acquisition
DC	Direct current
DCF	Distributed coordination function
DM	Deadline monotonic
DMS	Deadline monotonic scheduler
DOF	Degree of freedom
DSR	Dynamic source routing
EDF	Earliest deadline first
EOF	End of frame

FDMA	Frequency division multiplex access
FF	Field bus foundation
FIFO	First in, first out
FT	Frequency transmission
i.i.d	Independent and identically distributed
I/O	Input/output
IAE	Integral absolute error
IEC	International Electrotechnical Committee
IEEE	Institute of Electrical and Electronic Engineers
IFS	Interframe space
IMEP	Internet MANET Encapsulation Protocol
IP	Internet Protocol
ISO	International Standards Organization
KS	Knowledge source
LLF	Least laxity first
LMI	Linear matrix inequality
LQG	Linear quadratic Gaussian
LQR	Linear quadratic regulator
LTI	Linear time-invariant
LVDT	Linear variable differential transformer
MAC	Medium access control
MAGLEV	Magnetic levitation
MANET	Mobile ad hoc network
MAP	Manufacturing Automation Protocol
MAS	Multi-agent system
MATI	Maximum allowable transfer interval
MIMO	Multiple-input/multiple-output
MPR	Multipoint relays
NCS	Networked control system
NUT	Network update time
OLSR	Optimized link state routing
OSI	Open systems interconnection
OSPF	Open shortest path first
PC	Personal computer
PCF	Point coordination function
PDA	Personal digital assistant
PE	Priority exchange
PeE	Persistently exciting
PI	Proportional–integral
PID	Proportional–integral–derivative
PIV	Proportional–integral velocity
PROFIBUS	Process field bus
QoS	Quality of service
RIP	Routing Information Protocol
RM	Rate monotonic

RMS	Rate monotonic scheduler
RR	Round Robin
RTDS	Real-time distributed system
RTT	Round-trip Time
RVDT	Rotary variable differential transformer
SISO	Single-input/single-output
SLPMANET	Service Location Protocol to work in MANET
SOF	Start of frame
SOM	Self-organizing map
TC	Topology control
TCP	Transmission Control Protocol
TDMA	Time division multiplex access
TKS	Task knowledge structure
TOD	Try to discard
TORA	Temporally Ordered Routing Algorithm
TSK	Takagi–Sugeno–Kang
UDP	User Datagram Protocol
VLSI	Very large-scale integration
WCET	Worst case execution time
WLAN	Wireless local area network
WorldFIP	World Factory Instrumentation Protocol
WSDL	Web Services Description Language
XML	eXtensible Markup Language

Chapter 1
Introduction to Networked Control Systems

Abstract This chapter introduces a brief description of Networked Control Systems. A formal review of this book is given, describing the key issues within each chapter. A review of the strategies present in the literature is made to study and compensate for the network imperfections presenting three main methodologies. The control methodology focuses on generating control signals that counteract the effects of the network imperfections through modelling its dynamics or considering them as uncertainties. The communication methodology aims to improve the transmission of information and minimise imperfections through the scheduling and synchronisation of the nodes present in the network as a function of the system performance. The co-design methodology considers increasing the advantages of the various methodologies with the purpose of increasing system performance and minimising the effects of network imperfections. It is also presented the time delay modelling in nondeterministic networks, the main imperfection of the network. Finally, the maximum allowed transfer interval term is described which is the maximum bound for time imperfections of the network.

1.1 Basics of Networked Control Systems

A Networked Control System (NCS) is a particular kind of a real-time distributed system (RTDS), composed of a set of nodes, interconnected by a network, able to develop a control process. A NCS is a dynamic control system where the control loops are closed through a communication network. The main feature of a NCS is that the control signal, the feedback signal or both are exchanged among the system's components in a network, interchange information packages through a network.

Main components of a control process are sensors, controllers and actuators, which rely on a real-time operating system and a communication network. NCSs have received increasing attention in the last few years, due to several advantages, such as an easy maintenance, a high reliability, less wiring, and low cost. Besides, the use of control network architectures, when properly applied, may improve efficiency, flexibility, and reliability, reducing installation, reconfiguration, maintenance in time, and costs.

H. Benítez-Pérez et al., *Control Strategies and Co-Design of Networked Control Systems*, Modeling and Optimization in Science and Technologies 13,
https://doi.org/10.1007/978-3-319-97044-8_1

Nevertheless, notice that the use of a single communication network for all control signals (e.g. control and feedback) modify the performance of the whole NCS. This is due because the network generates some imperfections in packet transmission, such as time delays, packet loss, etc. Furthermore, conventional standard results in analysis and design of digital control theory cannot be used, since many of its assumptions are not directly applicable to NCS, specifically considering time delays and packet loss. The use of common bus network topology introduces different forms of uncertainties between sensors, actuators, and controllers. Several approaches have been developed to tackle this problem, mainly based on the work of Halevi and Ray [1]. In such approaches, a group of nodes make use of the same network, producing a traffic load. If there is no coordination among the nodes, some nodes may transmit simultaneously, this generates time-delays, retransmissions, or worst, packet loss. This results in packets which are unable to fulfil their deadlines. In order to achieve overall objectives of all message send in a NCS, it is necessary for all nodes to properly exchange their own information through the network. Therefore, the mechanism to exchange information plays an important role in the stability and performance of the NCS implemented over a communication network.

In a NCS, the main problems that degrade its performance are network-induced delays and packet loss. Time delays can be constant, time-varying or even random; those depend on the scheduler, network type, architecture, operating systems, etc. [2]. When a NCS uses a reliable network, it is possible to compensate for time delays, lower or larger than the sampling period obtaining their bounds. However, when a NCS uses an unreliable network, it is more difficult to compensate for unknown time delays and packet loss, and these reduce the efficiency of NCS.

NCSs have been studied for the last 20 years, using several assumptions regarding time delays modelling, or uncertainties accomplishment. In this book, some network imperfections are modelled, in order to be coupled with a control law design. Several strategies are reviewed, such as fuzzy approximation, frequency communication modelling, or dynamic scheduling procedure. All of them trigger certain threshold onto dynamic systems models. However, the modelling procedure for NCSs is a challenge, since there are several non-linear situations, like saturation, unknown time delays, dead-zones, etc., with which it is necessary to deal. Other static strategies, like digital control, are unable to produce a feasible result since time delays are variable.

As previously stated, time delays are dynamic as a result of computer network interaction. Therefore, to study such behaviour provides a good approximation of how time delays behave in several situations, like overflowing, fault appearance, packet loss, or undersampling time delays.

NCS, since its beginning, employed classic control techniques such as linear control, adaptable control, robust control, and so on, with some modifications to compensate the imperfections due to the network [3]. Such techniques, as it is common in classic control theory, are divided into techniques in the continuous dominion [4] and techniques in the discrete domain [5–7]. Fridman and Shaked [8] introduced a model transformation for the delay-dependent stability of systems with time-varying delays in terms of LMIs. Their model also refined results from delay-dependent $H\infty$ control and extended them to the case of time-varying delays.

Although the field of NCS is still relatively new, many research approaches have been developed in it. Most of them work on a single imperfection due to the network. Results present in the literature should be extended and integrated to simultaneously obtain a reference framework and study imperfections due to the network. Here, a revision of the literature around NCS is presented, focusing on methods that compensate several imperfections, and distinguishing design methodologies. Some of the techniques only compensate a single imperfection, such as time delays or packet loss as the most common ones, since both of them highly degrade the system performance. Later, some methods that tackle the combination of several imperfections are introduced, setting their advantages and liabilities.

The later techniques to design NCSs leave aside concepts established by communication networks, by supposing some features due to the network, like for example the maximum delay, its derivative, or the packet loss, as well as the assumption about the execution of strictly periodic nodes, and the synchronisation among them. Almost every imperfection arising from the process itself as plant delays [9].

In the last few years, the concept of co-design has been used to name the application of multiple knowledge areas together, aiming for solving a common problem. The most common co-design technique for NCS is based on control-communication, whose objective is to design a NCS through a controller observing network features, and a scheduler considering control aspects for the transmission between nodes.

As part of a co-design approach, it is especially important to take into consideration the multiple imperfections due to the network, not only for the controller design but also for the scheduler. For example, reducing the quantization error (by transmitting packets with more bits) normally results in longer time delays that have to be compensated, in the network by the scheduler, or in the process through a controller. Performing co-design assumes compensating the NCS as a whole (process, controller, scheduler), in an integral way, which makes necessary the use of tools to obtain quantitative information from the control system and the network.

1.2 Sources of Uncertainties

In NCS, the uncertainties induced by the network are categorised into five types:

1. Quantization error
2. Packet loss
3. Variant sampling interval/period
4. Communication delays
5. Communication limitations.

The presence of any of these uncertainties may degrade the performance of the system, even reaching instability. Based on this, it is important then to understand how these phenomena influence the stability of the system and its performance. Unfortunately, most of the available literature focuses only on a subset of them, while ignoring the rest. For example, systematic methods analyse the stability of NCSs considering only some imperfection. Quantization errors, for instance, are studied in [10–16].

In practice, almost all uncertainties of the network are present in a NCS. This makes necessary methods for analysis and synthesis that include most, if not all of these uncertainties, to compensate them. Very few results have been reported that integrate compensation for multiple uncertainties.

Modelling the network imperfections or modelling the network behaviour with imperfections, in several cases, is a prerequisite for control design, these models depend on the kind of network, the used protocol, and so on. The most common classification of time delays is proposed by Nilsson [2]. Three different models are discussed: constant time delay, random time delay dependent on previous transmissions, and random time delay with probability distribution dependent on Markov chains.

However, the models of independent random time delay do not capture the effect of strong package traffic in the network (also refer to consumed bandwidth), which sometimes follows correlations between random time delays, where a time delay value depends on previous time delays. If the network experiences strong traffic, it is very probable that all packets suffer long transmission time delays, until the network load diminishes. If the network load considerably varies, a solution is to model it using time delay distributions by a Markov chain [2].

For modelling a time delay in a network with strong and variant traffic, the model of the network may need to have a state unless the variation is unbounded. The effect of variant traffic could be modelled using a Markov chain, which performs a transition each time there is a transference on the communication network, and thus, proposing probability distributions for the time delays between sensor-controller (τ_{sc}), and controller-actuator (τ_{ca}) at each state.

One common strategy is the queue methodology, which uses observers and predictors to compensate the time delays, and ensures invariance through time of the NCS. Using the queue theory forces the random time delays to present a deterministic (constant) behaviour. The method presented by Luck and Ray [17] uses an observer to estimate the states of the plant, and a predictor to compute the predictive control based on measurements of past outputs. The control and the measurements of past outputs are stored as FIFO (First Input—First Output) queues in buffers. These are placed before and after the controller within the control loop. First, the past measurements are used for estimating the states of the plant in $k+1$, where k is the size of the buffer between sensor and observer. Then, using other estimations, the state of the plant is predicted at $k+l$, where l is the size of the buffer after the controller. The predictive control signal $u(k+l)$ is obtained and stored in the buffer. The observer and predictor are based on the dynamic model, and thus, the performance of the system is highly dependent on the precision of the model.

1.3 Communication Methodologies

Communication methodologies make use of computer science to design the network for the NCS and generating scheduling algorithms for the nodes and the network, aiming for maintaining the stability of the NCS. There are two main approaches: One characterises and designs the communication network [18–21]; and another design scheduling algorithms for the packet transmission through the network [22–25]. An objective of this approach is to keep the structure of the NCS, by establishing an Ethernet network without modification, and focussing on the scheduling methods.

Regarding scheduling, the pioneering work of Hong [22] proposes a scheduling algorithm to determine the sampling times by using the concept of a transmission window. Sampled data from the NCS components share a limited number of windows, so the performance requirements of each control loop are satisfied, but the utilisation of the network resources considerably increase. This methodology establishes time delays within a boundary, and avoiding packet collisions, applying to Polling and Token Passing systems.

Later, Pegden et al. [23] present an algorithm to assign bandwidth, applicable to the CAN protocol for NCS, using CSMA/NBA (Carrier Sense Multiple Access with Non-destructive Bitwise Arbitration) [26]. Each node watches the state of the bus before data transmission. Nodes delay data transmission until the bus is free. When the bus is free, any node may start transmission. If more than one node starts transmission at the same time, the bus enters in conflict, which requires solving a comparison of their identifiers. The algorithm to assign bandwidth satisfies the time delay requirements for real-time control and data for events, maximising the use of network bandwidth, and it is designed offline. It also presents the liability of being incapable of dynamically modify the assignment of data transmission, while the control system executes. Another liability is that it only takes into consideration time delays smaller than the sampling period.

Scheduling methodology provides that control actions are carried through events and no time, using the dynamics of the plant as triggers. Tarn and Xi et al. [27] propose a methodology based on events that use the movement of the system as a reference, instead of deciding the next sampling-action instant through time [28]. Such the action is started by an event, triggered by the system state or error. The method maps the time-space in the event-space, and thus, the stability of the system does not depend on time but the state's position or the measured error. Hence, the induced delays by the network are smaller than the response times of the process, and they do not destabilise the system, nevertheless they maybe accumulative. However, if the delays are too large or there is packet loss, the performance of the system degrades.

Queue methodology makes use of observers and predictors for compensating delays, guaranteeing time invariance of the NCS through time. The methodology uses queue theory to force that random delays present a deterministic and constant behaviour. The method by Luck and Ray [17] uses an observer to estimate the states of the plant, and a predictor to obtain the predictive control, based on previous output measurements. The control and previous output measurements are stored as FIFO queues in buffers. These are placed before and after the controller in the closed loop. First, previous output measurements are used to estimate the states of the plant, where it is the register size between sensor and observer. Next, using previous estimations, the state of the plant is predicted. The predictive control signal is obtained and stored in the register, the observer and predictor are based on a model, and thus, the performance of the system is highly dependent on how precise is the model. Chan and Ozguner et al. [29] used to jump systems to design a static controller for a time-invariant system with a FIFO queue and the known maximum buffer size and the upper bound of random delay in the communication link. More recently, Filipovic [30] used to finite dimensional discrete-time jump linear systems with finite state Markov chains but a non-linear model of a queue in the sensor, the feedback linearization is used to linearize the queue. It means that in the control loop there are two controllers. The first one performs feedback linearization of the buffer and the second one is the main feedback controller.

Wang and Liu's book [31] presents a review of the NCS systems in the past decade. It also presents a review of several scheduling protocols using in NCS like try-one-discard (TOD), Round-Robin (RR) persistently exciting (PE) constant penalty TOD (CP-TOD) etc., its advantages and disadvantages, finally presents an extensive stability analysis, other recient reviews are presented in [32].

1.4 Control Methodologies

The area of communications attempts to minimise, compensate, or eliminate the delays induced by the network, by an ordered assignment of transmission and maximising the resource utilisation. In the area of control, the intention is to incorporate the imperfections in the network model, and modifying the control techniques established for digital control and continuous control. The methods comprise optimal control, robust control, adaptive control, fuzzy control, stochastic control, and so on.

A common method is based on obtaining an augmented model in discrete time, which incorporates time delays in the network as a state of the system [1]. Also, time delays are considered as stochastic, and the objective is to design an optimal control [2]. Another recurrent approach is to suppose the delay as a perturbation of the system and compensate using control techniques to deal with perturbations [33]. The same happens using robust control for designing control laws that are not very sensitive to delays [36]. Adapting the control signal regarding the effect on the network is a method employed by fuzzy control and adaptable control, by modifying the controller gains.

For a review of other control methodologies can refer to Wang and Lius book [31] and introductions from [34, 35]. It presents diverse strategies such as predictive, $h\infty$, fuzzy, multi-agent, Smith predictor and robust polynomial. The fuzzy control systems presented are developed in the continuous and discrete domain, one showed a h∞ control for TSK (Takagi-Sugeno-Kang) systems with time delays, assuming synchronization of the nodes and only sensor-controller communication. The second uses a TSK model that considers the time delays and dropouts, modelling the imperfections as stochastic events. Both present simulation examples to prove their effectiveness. Also, Bemporad et al. book [37] presents a good review of the design characteristics for distributed control methodologies, from the selection and configuration of the network to the generation of middleware to minimize time and effort in the application. Performs a review of the modelling of wireless networks and its various topologies for NCS, An NCS review based on events is presented for its application in control, estimation and optimization.

The description of relevant methods and their evolution are described in the following paragraphs.

Halevi and Ray [1] propose an augmented method, augmenting the state space in the discrete-time model. Network delays are handled by the controller, using previously applied control signals as internal states for calculating a new control signal at the kth time. This generates a new model with the original states of the system and the previous inputs as new states. The stability for periodic delays is proven based on the eigenvalues of the transition matrix obtained of the augmented system.

Nilsson [2] develops an optimal control method for NCS. The delay is assumed as random, but smaller than the sampling period. Later Lincoln [38] extends the method for delays larger than the sampling period. The basic idea is to present the problem as a LQG (Linear Quadratic Gaussian) system. Dynamic systems are given in the space state, and the gain of the optimal controller is solved as a LQG problem, using dynamic programming. Solving this problem requires past delay information and the complete states.

The perturbation method, proposed by Walsh et al. [33], takes into consideration the difference between the actual output and the most recent transmitted output values as a perturbation of the system, searching the error boundaries. Stability is obtained using the Lyapunov method on the error dynamics. Several assumptions are established, including error-free communications, fast sampling, and noiseless observations. However, the process and controller may be non-linear and time variant. In this method, the network only sends from sensor to controller, and there is no exchange between controller and actuator. In [39], the networked control system is supposed for singularly perturbed systems with time-varying transmission times. The slow and fast systems of singularly perturbed systems are used to approximate the state behaviour. Under the condition that their systems are Hurwitz stable, the whole systems via the network control are robust practically stable if the perturbation parameter is sufficiently small.

In the robust control method, proposed by Goktas et al. [36], delays induced by the network are managed as perturbations of the nominal system. The controller design is carried out in the frequency domain, using robust control theory. The major advantage of this method is that it is not necessary to know beforehand the precision of the delay distribution. Network delays are supposed bounded, and they are modelled as simultaneous multiplicative perturbations. Thus, the $H\infty$ design synthesis can be applied, choosing certain multiplicative weights for uncertainties, such that the NCS would be not very sensitive to known delays.

The fuzzy control method takes advantage of fuzzy theory to update the gains of the controller based on the error signal between the reference and the actual output of the system and the controller output [40]. Actually, the method updates the gain of a PI (Proportional-Integral) controller based on the error and the controller output. The design includes member functions online and offline for optimisation. Zhang and Li et al. [41] obtain a delay for maximising the cross-correlation function of the output of the plant, estimated without delay, and the measured output. This estimation scheme uses a fuzzy Smith predictor that compensates the constant delay, and the fuzzy controller is used for adapting the changing parameters of the system.

The adaptive control is used for the modification of the system parameters provided counteract its dynamic changes. Thus optimising the control signals to each system behaviour. The adaptive control method for NCS is based on the ability to measure the traffic conditions of the network, adapting the controller parameters. In this method, the controller is able to sense and update the conditions of the Quality of Service (QoS) in the network. If the QoS requirements are not satisfied, the parameters of the controller are tuned, aiming to improve the performance as possible [42].

Packet loss is another network imperfection to highly take into consideration. In general literature, packet loss is modelled using two processes: a random Bernoulli process, or a typical Markov chain process. The results of the approaches that apply random Bernoulli assume that lost packets are independent and identically distributed (i.i.d.) [33, 43], while the other approaches that use Markov chains suppose that lost packets are streams, and occur in accordance with a Markov chain [44–46]. Most of the existing results design filters within the NCS to estimate the states of the system, with the feature of establishing an input zero, this is, when measurements are lost. With this configuration, the estimation with packet loss is called estimation with intermittent observations or loses [33, 43, 45, 46].

To estimate the system states with intermittent observations and to know the system behaviour with packet loss, to adapt a Kalman filter for the estimation is one of the most popular and useful methods for filtering problems. It has the disadvantages of supposing the knowledge of a precise model of the system and that the statistical information of external noise is already known. In [43], it is proposed the design of a Kalman filter using both packet loss models, Bernoulli i.i.d. and Gilbert-Elliot model (developed with discrete Markov chains with two states). Another result that employs a Kalman filter to estimate the states for linear, stochastic systems [33] takes into consideration the Bernoulli approach to model packet loss, as well as taking into consideration system uncertainties. Smith Y Seiler [45] employ a Kalman filter variant in simple time to estimate the states of the system, considering a Markov chains model for the lost packets. In [46] it is employed a discrete Kalman filter for packet modelling, using a binary probability distribution. The analysis and design of the filter provide a boundary in the arrival rate of measurements.

Another methodology for estimation is using filters, that provide a guarantee of noise attenuation and robustness over uncertainties of the model [47]. The filters for NCS have recently appeared [23, 47–49]. In [23], the filter is designed for a class of stochastic NCS with consecutive packet loss, separating the loss rate for sensor-controller communication and controller-actuator communication.

Regarding optimal controllers, Gupta et al. [50] developed a LQG controller for a link with a single packet lost between sensor and controller. It is assumed a linear process in discrete time, and without network induced delays within the control loop. The controller does not require information of the statistical model of the packet loss event. Imer et al. [51] and Sinopoli et al. [7] develop a LQG controller over unreliable communication links, comparing the performance of protocols such as TCP (Transmission Control Protocol) and UDP (User Datagram Protocol). Both studies model packet loss between sensor-controller link and controller-actuator link as Bernoulli i.i.d (Independent and Identically Distributed).

1.5 Co-Design Methodologies

Codesign has the objective of mixing the advantages of two or more knowledge areas and minimise the liabilities of such areas when presented separately, aiming to adequately resolve a common problem. In NCS, control-communication codesign commonly integrates feedback control and real-time communications. It is based on the principle that system performance depends on the design of control algorithms, and also relies on the scheduling performance of the shared communication resources. Unfortunately, NCS design is very frequently based on the principle of separation of concerns [52]. This principle assumes that feedback controllers can be modelled and implemented as periodic tasks with a fixed period, and known consumption (i.e. Worst Case Execution Time or WCET), and a rigid deadline. These assumptions are amply adopted by the control community and developed by sampling systems theory. Even though these assumptions allow the control community to focus on its own problems, without concerning about how scheduling is performed in the nodes and how packet in the network are carried out, as well as the communication community is not worried about understanding how scheduling impacts on system performance.

The control task does not always make use of the available computing resources in an optimal way, and the assumptions of the model of simple tasks are too restrictive regarding the characteristics of many control systems. For instance, many deadlines are not always rigid, that is, many practical control systems may occasionally tolerate losing a deadline.

To confront resource limitation in a NCS, co-design is necessary at different levels, for example, hardware/software codesign, or mechanic/electric codesign. Here, co-design is used at the control/communication level. This methodology can be divided into two categories: feedback control of computational systems and real-time control.

The category of feedback control of computational systems is also known as feedback scheduling [52]. The basic idea is dealing with the scheduling problem as a feedback control problem. A feedback control loop is introduced within the resource manager of the computer systems. The objective of feedback scheduling is increasing flexibility regarding the uncertainties in resource utilisation. Instead of pre-assigning resources based on an offline analysis, resources are online, dynamically assigned, based on the feedback from the utilisation of the actual resource and the execution times of control tasks.

Xia et al. [53] design a fuzzy feedback scheduling. Using a control loop and an outer feedback loop to implement the scheduling. The scheduler is to dynamically regulate the sampling periods to achieve a desired level of utilisation. The timing attributes of all non-control tasks cannot be changed by the feedback scheduler. Another work in this field is the fuzzy feedback scheduling based on output jitter, which is formulated in [54]. The output jitter is the controlled variable, and the CPU utilisation and jitter are the control inputs. So, when the system resource changes, the result is that some key periodic tasks still meet an expected jitter range regulating only the task periods.

The second category of codesign here focuses on real-time control. Branisky et al. [55] were among the first to establish that scheduling and control systems cannot be separately designed. They work on regulation problems of the plant as an optimal scheduling problem, observing the limitations of RM (Rate Monotonic) scheduling [56], as well as the limitations of regulating the NCS, taking into consideration network induced delays, packet loss, and multiple packet transmissions.

Branisky et al. [55] consider a set of NCSs with linear plants, transmitting only from the sensor to the controller, with a period equal to the deadline, and with a consumption time. Applying the RM scheduling algorithm, static priorities are assigned to each controlled plant. One plant with faster dynamics has a higher priority, so it has a higher transmission rate than a plant with slower dynamics. The set of N tasks is feasible if it accomplishes that the utilisation factor of the network U is less than 1.

For the scheduling optimisation, it is assumed that each NCS is associated with a measured performance function, giving the control cost as a function of the transmission period. The selection of the performance function is critical for the optimisation problem. Normally, a quadratic or exponential cost is chosen. As a result, this work analyses the percentage of allowed lost packets that secure the performance of a NCS. The NCS with lost packets is modelled as a dynamic asynchronous system. It is assumed that the sampling period is constant and that the NCS is able to tolerate a certain amount of lost feedback data. Even though the concept of co-design of a NCS is used, the method has some liabilities because the data transmission is considered only between the sensors and the controller, and a constant delay is supposed for the analysis of packet loss.

Sename et al. [57] make use of the concept of feedback control of computational resources for scheduling CPU (Central Processing Unit) resources and workload. They propose two schemes for the management of systems with a control task or multitask. The design of the control system takes into consideration unknown delays, given that time uncertainties are unavoidable. Likewise, they present a new method of control design using state feedback for systems with delays in discrete time as a LMI formulation. For the feedback scheduling, they consider communication delays mainly as the input/output latency. The objective is online tuning the sampling period of the controller, to accomplish with the requirements of the computational resources. They use discrete-time systems, in which the communication delay is the sum of the network-induced delay and the computational cost delay, for calculating the control input.

The presence of any of the five network imperfections can degrade the system performance, even reaching instability. Based on this, it is important to understand how these phenomena influence on the system stability and performance. Unfortunately, most of the available literature focuses on some of these imperfections, while disregarding then others. For example, systematic methods analyse NCS stability considering only a single imperfection. Quantization errors are studied in [10–16]; packet loss in [7, 50, 51, 58, 59]; variant interval/period sampling in [60–67], respectively; and the communication limitations in [68–71].

In practice, almost all network imperfection phenomena are present in a NCS, which implies the need of analysis and synthesis methods that include most, if not all, of these imperfections, to compensate them. Few results have been reported by integrating multiple imperfections. Some approaches that consider two types of imperfections are: [72–74], which focus on packet loss (2) and communication delays (4); [33, 48, 75–80] consider variant interval/period sampling (3) and communication limitations (5); [81] deals with quantization error (1) and communication delays (4).

Approaches that tackle three imperfections are [82], that deals with quantization error (1), variant interval/period sampling (3), and communication limitations (5); [83–85] work on packet loss (2), variant interval/period sampling (3), and communication delays (4); [86] focus on packet loss (2), variant interval/period sampling (3), and communication limitations (5); [87] incorporates quantization error (1), packet loss (2), and communication delays (4); and [88] deals with variant interval/period sampling (3), communication delays (4) and communication limitations (5).

Besides, some of the methods that study imperfections such as variant interval/period sampling (3) and communication delays (4), are able to study packet loss (2) as an extension. Chaillet and Bicchi [88] use a compensation method of the delay sending a larger control packet to the plant, which contains not only a control value for a particular time instant but also includes a control signal valid for a future near time horizon. The control scheme provides boundaries about the tolerable transmission delay and interval, such that the stability of NCS is guaranteed.

1.6 Time Delays Modelling

The nature of time delays is quite diverse, and commonly, inherent to the dynamic behaviour of the observed system. It may be either due to electromagnetic effects or due to mechanical sources. Hence, the nature of time delays has been the object of study of a vast community which, through the years, have produced several interesting results. Recently, time delays have been reviewed in terms of computing dynamics, like processor operation or network communications. In any case, time delays are able to be modelled as a perturbation of the system representation, or through stochastic differential equations [2, 48, 89, 90].

The main source of time delays is the non-linear behaviour as a result of the activity of operating systems and local scheduling management, as well as communicating data between processors. Different effects, like uncertainties, packet loss, long time delays, among others, are reviewed later in this book. Regarding control laws, several strategies for managing time delays have been studied by different research groups, like for example, the group of the University of Lund [91].

In general, time delays are measurable and bounded according to a real-time scheduling algorithm. For instance, the Earliest-Deadline-First (EDF) algorithm is a very well-known algorithm used for scheduling tasks. Using this algorithm, the schedulability of a system can be ensured, meeting the requirements of many case studies. Using EDF, tasks that need the common resource (generally, the communication network) are executed with the most restricted time deadline. This provides no time for unexpected and catastrophic time delays, due to a scheduling action. Hence, EDF implies a more efficient exploitation of computational resources and a higher responsiveness to aperiodic activities. Of course, there are other algorithms capable of obtaining the same goal, although they are commonly more complex. A deeper explanation of this algorithm is described in Sect. 5.3.

The scheduling of tasks, which execute as part of a NCS, has been widely studied, proposing several algorithms to achieve the highest level of resource utilisation, while minimising the effect of time delays [89, 92, 93]. A strategy that has proven to obtain an appropriate level in both, quality service and quality control, is the reconfiguration of states [94].

Besides, several authors define the relationship between control performance and sampling period, establishing a performance degradation point. Such a point is called "period" in digital control. From this, there are two degradation points, namely b and c, which correspond to two sampling periods, p_b and p_c, where the system operates in valid conditions within this range, i.e., the system presents an acceptable degradation due to sampling. Hence, in order to choose an appropriate sampling rate, several authors have searched different approaches. For example, [92] provides a clear way to choose a sampling period to minimise the effect of delays.

Testbeds of systems with time delays can apply algorithms in a controlled manner to observe their behaviour between processors and processes. Throughout the rest of this book, three testbeds are used to expose how time delays may be modelled and controlled: an LVDT Motor, a Magnetic Levitator system, and a Helicopter model.

Computer networks into the NCS introduce several problems unexpected in conventional control systems, for instance, packet loss and time-delays. Such effects occur simultaneously and influence the properties of the system or even lead to instability. Time delays have several reasons: system, computations and in a larger proportion in communications. Depending on the type of network used for a NCS, communication delays are commonly non-deterministic, generating uncertainties, which have to be taken into consideration when designing controllers.

There are some approaches to modelling time delays in NCS presented in the literature. For example, [95] take into consideration both the network-induced time

Table 1.1 First example for priority exchange

Name	Consumption (in units)	Period (in units)
Task 1	2	9
Task 2	1	9
Task 3	2	10
Server	1	6

delay and the time delay in the plant for proposing a controller design method, by using a time delay-dependent approach. An appropriate Lyapunov functional candidate is used to obtain a memoryless feedback controller. In this book, this is derived by solving a set of Linear Matrix Inequalities (LMIs) for constructing an $H\infty$ state feedback controller. Recently, Fridman and Shaked [8] introduce a new (descriptor) model transformation for delay-dependent stability, for systems with time-varying delays, in terms of LMIs. They also refine recent results on delay-dependent $H\infty$ control and extend them to the case of time-varying delays. Based on this review, a model is defined here, which integrates the time delays for a class of nonlinear system, therefore, a Fuzzy Control is presented for NCSs [96, 97], considering time delay induced by the network, as result of online reconfiguration. The stability analysis is revised as well.

Time delays are supervised as follows:

$$\varepsilon_1 \rightarrow \varepsilon_n / T_1 \rightarrow T_n \tag{1.1}$$

providing a strict priority from consumption times ($\varepsilon*$) and the related periods ($T*$) where priority is given as the well-known EDF algorithm [98], which establishes the process with the closest deadline has the most important priority.

To do so, several algorithms like Priority Exchange (PE) algorithms are used to manage spare time from EDF algorithm. The PE algorithm [99] uses a virtual server that deploys a periodic task with the highest priority, in order to provide enough computing resources for aperiodic tasks. This simple procedure provides a proximity, deterministic, and dynamic behaviour within the group of included processes. In this case, time delays can be deterministic and bounded. As an example, let us consider a group of tasks, as shown in Table 1.1.

The following task ordering is shown in Fig. 1.1, using a PE algorithm, where clearly time delays appear. Now, from this resulting different ordering, tiny time delays are given for two scenarios, as shown in Fig. 1.2.

These two scenarios present two different local time delays that need to be taken into account beforehand, in order to settle the related delays, according to scheduling approach and control design. These time delays can be expressed in terms of local relations among dynamical systems. These relations are the actual and possible time delays, bounded as marked by a limit of possible and current scenarios. Then, time delays may be expressed as local summations, with a high degree of certainty, for each specific scenario.

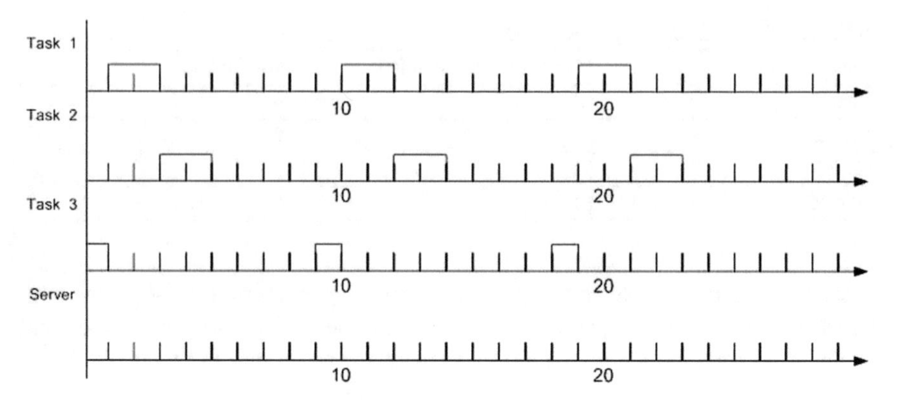

Fig. 1.1 Task ordering using a PE algorithm

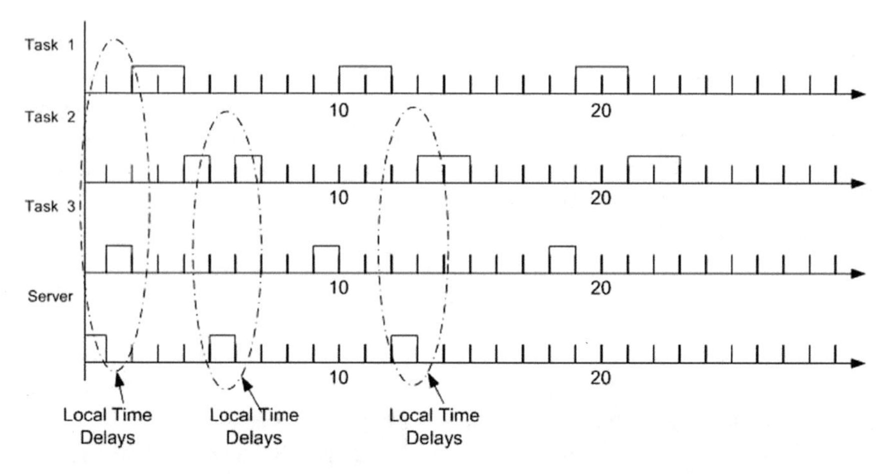

Fig. 1.2 Tiny time delays for two scenarios

A distributed system is loosely coupled, and thus, it has a high cost to know its global state, given the communication among its elements. The scheduling in each agent allows to achieve the time restrictions, and takes decisions that control the access to resources in an independent way; however, this may carry out decisions that, globally, are not at all certain.

In this design stage of distributed control systems, also known as NCS, it is considered as an important element of analysis to determine the range of sampling periods in which the system presents an acceptable performance.

Nowadays, the trend in systems that integrate computing, communication, and control, make use of control architectures with a common communication network. This improves the efficiency, flexibility, and reliability of distributed systems. Nevertheless, this architecture implies several kinds of time delays between sensors, controllers, and actuators, due to they share a communication media, as well as due to the communication process. Cervin et al. [100] discuss the possibility of analysing

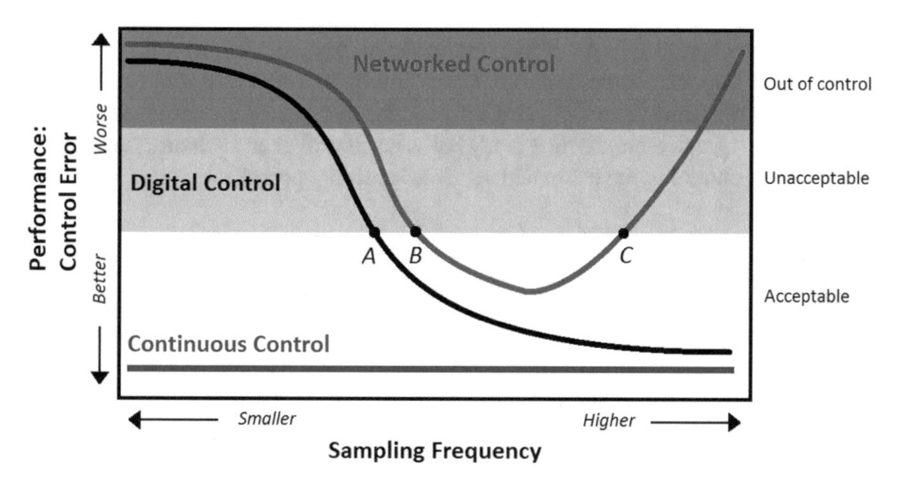

Fig. 1.3 Performance comparison of continuous control, digital control, and networked control cases (Modified from [26])

the performance of controller under the effect of modifying the sampling intervals and the time delays of communication.

The analysis is fundamentally based on the performance diagram in Fig. 1.3, derived from comparing the control performance against the sampling frequency. This diagram contrasts the three types of system control: continuous control, digital control, and NCS. Further, it provides several points of performance degradation. Continuous control is considered as the ideal, in which a good performance of the system is held at every moment, given that the performance does not depend on any sampling frequency. In digital control, the sampling frequency only is limited for the performance of CPU and it has an ideal sampling frequency (Point A) defined by the system dynamics. But in the case of networked control, it has two critical points. The point B is defined by the system dynamics and the point C is defined by constraint communications because with a higher sampling frequency the network generates more imperfections.

The exponential distribution is used supposing that the distance between nodes is relatively short (just a router) and the Gaussian distance for relatively long distances (multiple routers). For short distances, the time delay distributions can be divided into two parts: one constant and one variable. The constant part is defined due to the communication time delay provided by the physical length of the wiring and the processing time of the involved nodes, while the variable part is mostly affected by the stack of packages in elements such as switches and routers [99].

Having defined time delays as the result of a scheduling and networking approach, as well as external events like faults (just local and bounded faults), several scenarios are potentially presented following this time delay behaviour. In fact, the number of scenarios is finite, since the combinatorial formation is bounded as well as the kind of time delay.

In this case, one-way time delays and package loss are taken as a variable sampling period ($\hat{h}_v$). For computing this in the controller node, first it is obtained the sensor-controller time delay (τ^{sc}) (Eq. 1.3) and the lost packages between sensor and controller (τ^{lsc}), and then, the time delay between controller and actuator is estimated ($\hat{\tau}^{ca}$), finally obtaining the estimated variable sampling period $\hat{h}_v$.

$$\hat{h}_v = \tau^{sc} + \widehat{\tau}^{ca} \tag{1.2}$$

The lost packages between sensor and controller (τ_i^{lsc}) are used to estimate the lost packages between controller and actuator, which are measured without information for calculating the controller signal, and that modify the estimate of the time delay between controller and actuator (τ^{ca}). Supposing there is a controller node with timestamps for package arrival from the sensor (c_i) and timestamp for package emission to the sensor node (s_i):

$$\begin{aligned} \tau_i^j &= \left(c_i - \overrightarrow{c}_{i-1}\right) - (s_i - s_{i-1}) \\ \tau^{sc} &= \tau_0^{sc} + \tau_i^j \\ \overrightarrow{c_i} &= c_i - \tau_i^j \end{aligned} \tag{1.3}$$

$$\tau_i^{lsc} = ((s_i - s_{i-1}) - T_i) \tag{1.4}$$

where τ_i^j is the jitter (time delay variation) that composes the current sensor-controller time delay (t^{sc}), along with the unknown initial time delay (t_0^{sc}), $\overrightarrow{c_i}$ is the execution time of the c_i controller without jitter, with s_i and c_i as the timestamps of the packets in the sensor and control nodes, respectively. t_i^{lsc} are the lost packets between sensor and controller at time i, and T_i is the sampling period at time i (Fig. 1.4). Once the time delay τ_i^j of the system is estimated, it is used to generate a control signal through a fuzzy controller, as presented in Chap. 4.

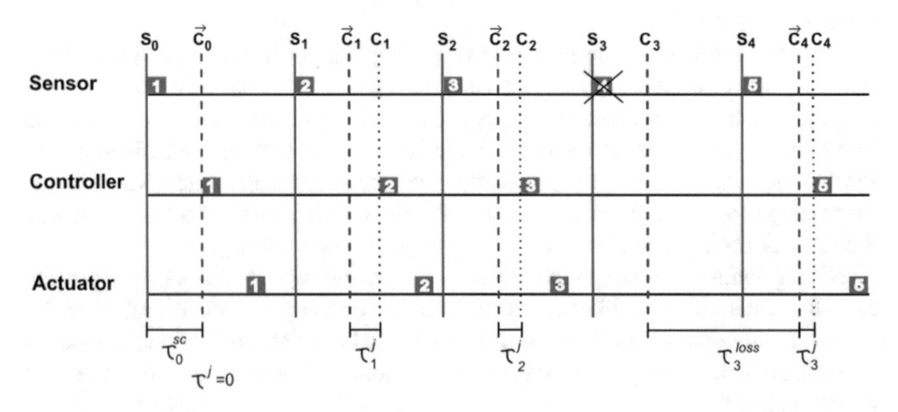

Fig. 1.4 Representation of time delays during communications

1.7 Maximum Allowable Transfer Interval

Maximum Allowable Transfer Interval (MATI) is a strategy to determine certain kind of intervals in terms of transmission, which is to bound the time delay. This is possible since the source of time delays is the communication in the network. MATI has been explored in [75], giving a clear condition for the worst case scenario, but in a dynamic strategy, and guarantee limit conditions, either using Lyapunov Krasovskii strategy [101], or TSK Fuzzy control.

In [33], Walsh introduced static and dynamic scheduling policies, but just for transmission from the sensor to controller in a continuous-time Linear Time Invariant (LTI) system. They introduce the notion of MATI as the longest time in which a sensor should transmit a datum. Therefore, Walsh derived Try-Once-Discarded (TOD) scheduling where the MATI constraint ensures at least one such transmission every T seconds. However, TOD does not guarantee that each node transmits once every p transmissions. In [67], Zhang extends [33], developing a theorem that ensures the negative nature of the Lyapunov derivative function, defined for a discrete-time LTI system at each sampling instant.

Regarding NCS, [2] also analyses several important facets of NCSs, by introducing models for the delays in NCS: first as a fixed delay, then as an independently random, and finally, like a Markov process. Optimal stochastic control theorems for NCSs are introduced, based on the independently random and Markovian delay models. Koubias [102] introduced static and dynamic scheduling policies for transmission of sensor data in a continuous-time LTI system. They introduce the notion of the maximum allowable transfer interval (MATI). This is the longest time after which a sensor should transmit a data. They also derive bounds of the MATI, such that the NCS is kept stable. This MATI ensures that the Lyapunov function of the system under consideration is strictly decreasing at all times [103]. This work extends the work in [102] developing a theorem which ensures the decrease of a Lyapunov function for a discrete-time LTI system at each sampling instant, by using two different bounds. These results are less conservative than those in [33] since here it is not required that the system Lyapunov function should be strictly decreasing at all time. Further, a number of different LMI tools for analysing and designing optimal switched NCSs are introduced. Alternatively, [99] takes into consideration both the network-induced delay and the time delay in the plant, and thus, introduces a controller design method, using a delay-dependent approach.

In a NCS, the main problems that degrade NCS performance are network-induced delays and packet loss. Time delays can be constant, time-varying or even random; those depend on the scheduler, network type, architecture, operating systems, etc. [2]. When a NCS uses a reliable network, it is possible to compensate for time delays lower or larger than the sampling period and obtain bounds. However, when a NCS uses an unreliable network, it is more difficult to compensate for the time delays and packet loss, and these reduce the efficiency of NCS. Therefore, analyse time delay and packet loss to develop an efficient approach and reduce its effect is critical for NCS with an unreliable network.

Nilsson has been a pioneer, analysing several important facets of NCSs [2]. He introduced models for fixed, independently random and Markovian time delays in NCSs. His paper introduced optimal stochastic control theorems for NCSs based upon the independently random and Markovian delay models. In [33], static and dynamic scheduling policies for transmission of sensor-controller data in continuous-time, linear time-invariant (LTI) systems are reviewed. Furthermore, the notion of the MATI was introduced, which is the longest time for that a sensor transmit a data with the warranty of a stable NCS. In [104], the work of Walsh was extended. He developed a theorem, which ensures the decrease of a Lyapunov function for a discrete-time LTI system at each sampling instant, using two different bounds; in the last two cases, only the communication between the sensor and controller is provided.

1.8 Organization of the Book

This book contains seven chapters. This first chapter presents an introduction to NCS, collecting of the most common methodologies in the NCS design, also shows some fundamental concepts for the development of this book, such as the time delay modelling and the definition of a maximum bound to consider that the system is stable. Chapter 2 presents a statistical model of the time delays generated within a non-deterministic communication network. Also, three NCS models are presented to incorporate the effects of network imperfections into the design, as well as an identification model for NCS. Chapter 3 presents a brief introduction to the modelling of distributed systems, considering their dynamic behaviour based on real-time representation, such as scheduling algorithms and relationship matrices. Chapter 4 presents the NCS design for each of the models presented in Chap. 2, generating the control and scheduling laws as well as a control-scheduling codesign that improve the system performance. Chapter 5 shows some approaches for designing mobile control applications, taking into account consensus and routing algorithms for decision making, as well as the delays generated by the makings decision. Chapter 6 provides two cases studies that are used to apply the control, scheduling and codesign methodologies presented throughout the previous chapters, providing the details of the effects of time delays, as well as the proper NCS design for each one of the challenging experimental benchmarks. Finally, Chap. 7 provides a summary of the contents of all previous chapters, in the form of a conclusion, and some aims for future work.

References

1. Halevy, Y., Ray, A.: Integrated communication and control systems: part I analysis. J. Dyn. Syst. Meas. Control **110**, 367–373 (1988)
2. Nilsson, J.: Real-time control systems with delays. Ph.D Thesis, Department of Automatic Control, Lund Institute of Technology, Lund, Sweden (1998)

3. Heemels, M., Van de Wouw, A.N., Nesic, D.: Networked control systems with communication constraints tradeoffs between transmission intervals, delays and performance. IEEE Trans. Autom. Control **55**(8) (2010)
4. Lincoln, B., Bernhardsson, B.: Optimal control over networks with long random delays. In: Proceedings International Symposium on Mathematical Theory of Networks and Systems, January 2000
5. Nian, X.: Stability of linear systems with time-varying delays: an Lyapunov functional approach. In: Proceedings of the American Control Conference, Denver, USA, June 2003
6. Pan, Y.J., Marquez, H.J., Chen, T.: Stabilization of remote control systems with unknown time varying delays by LMI techniques. Int. J. Control **79**(7), 752–763 (2006)
7. Sinopoli, B., Schenato, L., Franceschetti, M., Poolla, K., Sastry, S.: An LQG optimal linear controller for control systems with packet losses. In: Proceedings of the 44th IEEE Conference on Decision and Control, and the European Control Conference 2005, Sevilla, Spain, December 2005
8. Fridman, E., Shaked, U.: Delay-dependent stability and H∞ control: constant and time-varying delays. Int. J. Control **76–1**, 48–60 (2003)
9. Chen, J., Gu, K., Kharitonov, V.: Stability of Time-Delay Systems. Birkhauser, Boston (2002)
10. Brockett, R., Liberzon, D.: Quantized feedback stabilization of linear systems. IEEE Trans. Autom. Control **45**(7), 1279–1289 (2000)
11. Delchamps, D.F.: Stabilizing a linear system with quantized state feedback. IEEE Trans. Autom. Control **35**(8), 916–924 (1990)
12. Heemels, W.P.M.H., Gorter, R.J.A., van Zijl, A., Bosch, P.P.J.V.D., Weiland, S., Hendrix, W.H.A., Vonder, M.R.: Asynchronous measurement and control: a case study on motor synchronisation. Control Eng. Prac. **7**(12), 1467–1482 (1999)
13. Heemels, W.P.M.H., Siahaan, H., Juloski, A., Weiland, S.: Control of quantized linear systems: an optimal control approach. In: Proceedings of American Control Conference, Denver, CO, pp. 3502–3507 (2003)
14. Liberzon, D.: On stabilization of linear systems with limited information. IEEE Trans. Autom. Control **48**(2), 304–307 (2003)
15. Nair, G.N., Evans, R.J.: Stabilizability of stochastic linear systems with finite feedback data rates. SIAM J. Control Optim. **43**, 413–436 (2004)
16. Tatikonda, S., Mitter, S.K.: Control under communication constraints. IEEE Trans. Autom. Control **49**(7), 1056–1068 (2004)
17. Luck, R., Ray, A.: An observer-based compensator for distributed delays. Automatica **26–5**, 903–908 (1990)
18. Eker, J., Cervin, A.: Distributed wireless control using Bluetooth. In: Proceeding of IFAC Conference on New Technologies for Computer Control, Hong Kong, P.R. China, pp. 1–6, November 2001
19. Ji, K., Kim, W.J., Srivastava, A.: Internet-based real-time control architectures with time-delay/packet-loss compensation. Asian J. Control **9**(1) (2006)
20. Li, Q., Mills, D.L.: Jitter-based delay-boundary prediction of wide-area networks. IEEE/ACM Trans. Netw. **9**(5), 578–590 (2001)
21. Ploplys, N.J., Kawka, P.A., Alleyne, A.G.: Closed-loop control over wireless network. IEEE Control Syst. Mag. **24**(3), 58–71 (2004)
22. Hong, S.H.: Scheduling algorithm of data sampling times in the integrated communication and control systems. IEEE Trans. Control Syst. Technol. **3**(2) (1995)
23. Pegden, C.G., Shannon, R.E., Sadowski, R.P.: Introduction to Simulation Using SIMAN. McGraw Hill (1995)
24. Lu, C., Stankovic, J.A., Tao, G., Son, S.H.: Feedback control real-time scheduling: framework, modeling and algorithms. Real-Time Syst. **23**, 85–126 (2002)
25. Tanenbaum, A.S.: Computer Networks, 3rd ed. Prentice-Hall Inc. (1996)
26. Lian, F., Moyne, J., Tilbury, D.: Network design consideration for distributed control systems. IEEE Trans. Control Syst. Technol. **10**(2) (2002)

27. Tarn, T.J., Xi, N.: Planning and control of internet-based teleoperation. In: Proceedings of SPIE: Telemanipulator and Telepresence Technologies V, Boston, MA, vol. 35, no. 24, pp. 189–193 (1998)
28. Castillo-Gutierrez, O.O., Benitez, Perez H.: A Novel Technique to Enlarge the Maximum Allowable Delay Bound in Sampled-Data Systems. Congreso Nacional de Control Automatico, Mexico (2017)
29. Chan H., Ozguner U.: Optimal control of systems over a communication network with queue via a jump system approach. In: Proceedings of the IEEE Conference on Control Applications, pp. 1148–1153 (1995)
30. Filipovic, V.Z.: Robust control of systems over communication network. Sci. Tech. Rev. **57**(2), 24–30 (2007)
31. Wang, F.Y., Liu, D.: Networked Control Systems Theory and Applications. Springer, London (2008)
32. Benitez-Perez, H., Benitez-Perez, A., Ortega-Arjona, J.,Esquivel-Flores, O.: Networked control systems design consideringscheduling restrictions. Int. J. Adv. Fuzzy Syst. **2012**, 9(2012). Article ID 927878
33. Walsh, G.C., Beldiman, O., Bushnell, L.G.: Asymptotic behavior of nonlinear networked control systems. IEEE Trans. Autom. Control **46**(7), 1093–1097 (2001)
34. Quinones-Reyes, P., Benitez-Perez, H., Cardenas-Flores, F.,Ortega-Arjona, J.: Reconfigurable fuzzy Takagi Sugeno modelpredictive control networked control (Magnetic Levitation CaseStudy). Proc. Inst. Mech. Eng. Part I J. Syst. Control Eng.**224–8**, 1022–1032 (2010)
35. Benitez-Perez, H., Ortega-Arjona, J., Cardenas-Flores, F.,Quinones-Reyes, P.: Reconfiguration control strategy usingTakagi-Sugeno model predictive control for network controlsystems—a magnetic levitation case study. Proc. Inst. Mech. Eng.**224**(I8), 1022–1032 (2010)
36. Goktas, F., Smith, J. M., Bajcsy, R.: μ-Synthesis for distributed control systems with network-induced delays. In: Proceedings of the 35th IEEE Conference on Decision and Control, vol. 1, pp. 813–814. IEEE (1996)
37. Bemporad, A., Heemels, M., Johansson, M.: Networked Control Systems, vol. 406. Springer, Heidelberg (2010)
38. Lincoln, B.: Dynamic programming and time-varying delay systems. Ph.D. Thesis, Lund Institute of Technology (2003)
39. Yu, H., Zhang, B.: Stability of model-based networked control singularly perturbed systems with time-varying transmission times. In: 2016 Chinese Control and Decision Conference (CCDC), pp. 5722–5725. IEEE (2016)
40. Almutairi, N.B., Chow, M.Y., Tipsuwan, Y.: Network-based controlled dc motor with fuzzy compensation. In: Proceedings The 27th Annual Conference of the IEEE Industrial Electronics Society, Denver, USA, vol. 3, pp. 1844–1849 (2001)
41. Zhang, T., Li, Y.C.: A fuzzy smith control of time-varying delay systems based on time delay identification. In: Proceedings of the Second International Conference on Machine Learning and Cybernetics, November 2003
42. Tipsuwan, T., Chow, M.Y.: Network-based controller adaptation based on QoS negotiation and deterioration. In: Proceedings of the 27th Annual Conference of the IEEE Industrial Electronics Society, Denver, USA, vol. 3, pp. 1794–1799 (2001)
43. Lankes, S., Reke, M., Jabs, A.: A time-triggered Ethernet protocol for real-time CORBA. In: Proceedings of the 5th IEEE International Symposium on Object-Oriented Real-Time Distributed Computing, Washington, DC, USA, pp. 215–222, April 2002
44. Filipiak, J.: Modelling and Control of Dynamic Flows in Communication Networks. Springer (1988)
45. Aström, K.J., Wittenmark, B.: Computer-Controlled Systems: Theory and Design, 2nd edn. Prentice-Hall Inc., Englewood Cliffs (1990)
46. Liou, L.W., Ray, A.: Integrated communication and control systems: part III—Nonidentical sensor and controller sampling. J. Dyn. Syst. Measurement Control **112**, 357–364 (1990)

47. Tipsuwan, Y., Chow, M.Y.: Control methodologies in networked control systems. Control Eng. Pract. **11**(10), 1099–1111 (2003)
48. Walsh, G.C., Ye, H., Bushnell, L.G.: Stability analysis of networked control systems. IEEE Trans. Control Syst. Technol. **10**(3), 438–446 (2002)
49. Khalil, H.K.: Nonlinear Systems, 2nd edn. Prentice Hall, Upper Saddle River (1995)
50. Gupta, V., Spanos, D., Hassibi, B., Murria, M.R.: On LQG control across a stochastic packet-dropping link. In: Proceedings 2005 American Control Conference, Portland, USA, pp. 360–365, June 2005
51. Imer, O.C., Yuksel, S., Basar, T.: Optimal control of LTI systems over unreliable communication links. Automatica **42**(9), 1429–1439 (2006)
52. Arzen, K.E., Bernhardsson, B., Eker, J., Cervin, A, Nilsson, K., Persson, P., Sha, L.: Integrated control and scheduling. Technical Report ISRN LUTFD2/TFRT7586SE, Lund Institute of Technology, Sweden (1999)
53. Xia, F., Sun, Y.X., Tian, Y.C., Tade, M.O., Dong, J.X.: Fuzzy feedback scheduling of resource-constrained embedded control systems. Int. J. Innovative Comput. Inf. Control **5**, 311–321 (2005)
54. Wan, J., Li, D.: Fuzzy feedback scheduling algorithm based on output jitter in resource-constrained embedded systems. In: 2010 IEEE International Conference on Environmental Science and Computer Engineering, vol. 2, pp. 457–460 (2010)
55. Branisky, M.S., Phillips, S.M., Zhang, W.: Scheduling and feedback co-design for networked control systems. Proc. IEEE Conf. Decis. Control **2**, 1211–1217 (2002)
56. Liu, J.W.S.: Real-Time Systems. Prentice Hall (2000)
57. Sename, O., Simon, D., Robert, D.: Feedback scheduling for real-time control of systems with communication delays. In: Proceedings of the IEEE Conference on Emerging Technologies and Factory Automation (2003)
58. Sinopoli, B., Schenato, L., Franceschetti, M., Poolla, K., Jordan, M.I., Sastry, S.S.: Kalman filtering with intermittent observations. IEEE Trans. Autom. Control **49**(9), 1453–1464 (2004)
59. Smith, S.C., Seiler, P.: Estimation with lossy measurements: jump estimators for jump systems. IEEE Trans. Autom. Control **48**(12), 2163–2171 (2003)
60. Montestruque, L.A., Antsaklis, P.: Stability of model-based networked control systems with time-varying transmission times. IEEE Trans. Autom. Control **49**(9), 1562–1572 (2004)
61. Fujioka, H.: Stability analysis for a class of networked/embedded control systems: a discrete-time approach. In: Proceedings of American Control Conference, pp. 4997–5002 (2008)
62. Cloosterman, M., van de Wouw, N., Heemels, W., Nijmeijer, H.: Stability of networked control systems with uncertain time-varying delays. IEEE Trans. Autom. Control **54**(7), 1575–1580 (2009)
63. Gielen, R.H., Olaru, S., Lazar, M., Heemels, W.P.M.H., van de Wouw, N., Niculescu, S.I.: On polytopic inclusions as a modeling framework for systems with time-varying delays. Automatica **46**(3), 615–619 (2010)
64. Hetel, L., Daafouz, J., Iung, C.: Stabilization of arbitrary switched linear systems with unknown time-varying delays. IEEE Trans. Autom. Control **51**(10), 1668–1674 (2006)
65. Kao, C.Y., Lincoln, B.: Simple stability criteria for systems with time-varying delays. Automatica **40**, 1429–1434 (2004)
66. Naghshtabrizi, P., Hespanha, J.P., Teel, A.R.: Stability of delay impulsive systems with application to networked control systems. In: Proceedings of American Control Conference, New York, pp. 4899–4904 (2007)
67. Zhang, L., Shi, Y., Chen, T., Huang, B.: A new method for stabilization of networked control systems with random delays. IEEE Trans. Autom. Control **50**(8), 1177–1181 (2005)
68. Brockett, R.: Stabilization of motor networks. In: Proceedings of 34th IEEE Conference on Decision Control, vol. 2, pp. 1484–1488 (1995)
69. Dačić, D.B., Nešić, D.: Quadratic stabilization of linear networked control systems via simultaneous protocol and controller design. Automatica **43**, 1145–1155 (2007)
70. Hristu, D., Morgansen, K.: Limited communication control. Syst. Control Lett. **37**(4), 193–205 (1999)

71. Rehbinder, H., Sanfridson, M.: Scheduling of a limited communication channel for optimal control. Automatica **40**(3), 491–500 (2004)
72. Cloosterman, M., van de Wouw, N., Heemels, W., Nijmeijer, H.: Stabilization of networked control systems with large delays and packet dropouts. In: Proceedings of American Control Conference, pp. 4991–4996 (2008)
73. Gupta, V., Chung, T.H., Hassibi, B., Murray, R.M.: On a stochastic sensor selection algorithm with applications in sensor scheduling and sensor coverage. Automatica **42**(2), 251–260 (2006)
74. Matveev, A.S., Savkin, A.V.: The problem of state estimation via asynchronous communication channels with irregular transmission times. IEEE Trans. Autom. Control, **48**(4), 670–676 (2003)
75. Carnevale, D., Teel, A.R., Nesic, D.: A Lyapunov proof of improved maximum allowable transfer interval for networked control systems. IEEE Trans. Autom. Control **55**, 892–897 (2007)
76. Donkers, M.C.F., Hetel, L., Heemels, W.P.M.H., van de Wouw, N., Steinbuch, M.: Stability analysis of networked control systems using a switched linear systems approach. In: Hybrid Systems: Computation and Control. Lecture Notes in Computer Science, pp. 150–164. Springer, New York (2009)
77. Hetel, L., Daafouz, J., Iung, C.: Analysis and control of LTI and switched systems in digital loops via an event-based modeling. Int. J. Control **81**(7), 1125–1138 (2008)
78. Liberzon, D.: Quantization, time delays, and nonlinear stabilization. IEEE Trans. Autom. Control **51**(7), 1190–1195 (2006)
79. Nesic, D., Teel, A.R.: Input-to-state stability of networked control systems. Automatica **40**, 2121–2128 (2004)
80. van de Wouw, N., Naghshtabrizi, P., Cloosterman, M., Hespanha, J.P.: Tracking control for sampled-data systems with uncertain time-varying sampling intervals and delays. Int. J. Robot. Nonlin. Control **20**(4), 387–411 (2010)
81. Heemels, W.P.M.H., Nesic, D., Teel, A.R., van deWouw, N.: Networked and quantized control systems with communication delays. In: Proceedings of Joint IEEE Conference on Decision Control (CDC) 28th Chinese Control Conference, pp. 7929–7935 (2009)
82. Nesic, D., Liberzon, D.: A unified framework for design and analysis of networked and quantized control systems. IEEE Trans. Autom. Control **54**(4), 732–747 (2009)
83. Naghshtabrizi, P., Hespanha, J.P.: Designing an observer-based controller for a network control system. In: Proceedings of the 44th IEEE Conference on Decision and Control, and the European Control Conference 2005, Sevilla, Espaa, December 2005
84. Cloosterman, M., Hetel, L., van deWouw, N., Heemels, W., Daafouz, J., Nijmeijer, H.: Controller synthesis for networked control systems. Automatica **46**(10) (2010)
85. Naghshtabrizi, P., Hespanha, J.P.: Stability of network control systems with variable sampling and delays. In: Proceedings of the Annual Allerton Conference on Communication, Control, and Computing CD ROM (2006)
86. Nesic, D., Teel, A.R.: Input-output stability properties of networked control systems. In: IEEE Trans. Autom. Control **49**(10), 1650–1667 (2004)
87. Gao, H., Chen, T., Lam, J.: A new delay system approach to network-based control. Automatica **44**(1), 39–52 (2008)
88. Chaillet, A., Bicchi, A.: Delay compensation in packet-switchingnetworked controlled sytems. In: Proceedings of IEEE Conference on Decision Control, pp. 3620–3625 (2008)
89. Lian, F., Moyne, J., Tilbury, D.: Network architecture and communication modules for guaranteeing acceptable control and communication performance for networked multi-agent systems. IEEE Trans. Industr. Inf. **2–1**, 12–24 (2006)
90. Fridman, E.: Introduction to Time-Delay Systems. Birkhäuser (2014)
91. Cervin, A., Ohlin, M., Henriksson, D.: Simulation of networked control systems using truetime. In: Proceedings of the 3rd International Workshop on Networked Control Systems: Tolerant to Faults, Nancy, France (2007)

92. Lian, F., Moyne, J., Otanez, P., Tilbury, D., Moyne, J.: Design of sampling and transmission rates for achieving control and communication performance in networked multi-agent system In: Proceedings of American Control Conference, pp. 3329–3334, Denver, USA, June 4–6 (2003)
93. Branicky, M., Liberatore, V., Phillips, S.: Networked control system co-simulation for co-design. In: Proceedings of American Control Conference, Denver, USA, pp. 3341–3346, 4 June 2003
94. Benitez-Perez, H., Garcia-Nocetti, F.: Reconfigurable Distributed Control. Springer, Berlin (2005)
95. Zhu, X., Hua, C., Wang, S.: State feedback controller design of networked control systems with time delay in the plant. Int. J. Innov. Comput. Inf. Control **4**(2), 283–290 (2008)
96. Zhang, H., Yang, D., Chai, T.: Guaranteed cost networked control for T-S fuzzy systems with time delays. IEEE Trans. Syst. Man Cybern. C **37–2**, 160–172 (2007)
97. Tanaka, K., Wang, H.O.: Fuzzy Control Systems Design and Analysis: A Linear Matrix Inequality Approach. Wiley (2001)
98. Esquivel-Flores, O., Benitez-Perez, H.: Reconfiguracion dinamica de sistemas distribuidos en tiempo real basada en agentes. Revista Iberoamericana de Automatica e Informatica Industrial **9–3**, 300–313 (2012)
99. Mendez-Monroy, P.E., Benitez-Perez, H.: Supervisory fuzzy control for networked control systems. Int. J. Innov. Comput. Inf. Control Exp. Lett. **3–2**, 233–240 (2009)
100. Cervin, A., Henriksson, D., Lincoln, B., Eker, J., Arzen, K.-E.: How does control timing affect performance? Analysis and simulation of timing using jitterbug and truetime. IEEE Control Syst. **23**(3), 16–30 (2003)
101. Fridman, E., Niculescu, S.I.: On complete Lyapunov-Krasovskii functional techniques for uncertain systems with fast-varying delays. Int. J. Robust Nonlinear Control **18**(3), 364–374 (2008)
102. Koubias, S.A., Papadopoulos, G.D.: Modern fieldbus communication architectures for real-time industrial applications. Comput. Ind. **26**(3), 243–252 (1995)
103. Corkill, P.: Collaborating software: blackboard and multiagent systems and the future. Proc. Int. Lisp Conf. **3**, 123–138 (2003)
104. Zhang, W.: Stability analysis of networked control systems, Ph.D. Thesis Department of Electrical Engineering and Computer Science, Case Western Reserve University (2001)

Chapter 2
Modelling of Networked Control Systems

Abstract This chapter shows models for time delays and others network imperfections generated into NCS and how they are integrated into control, scheduling or codesign algorithms. First, a time delay model is presented using a generalized exponential distribution based function with data collect from non-deterministic networks. After, three NCS models are presented, each incorporates information about the network imperfections with the ultimate aim of generating a corrective action. We present models based on control, communication and codesign methodologies. Finally, a neuro-fuzzy identification is presented to model the system states and estimate the parameters of the NCS based on multi-sampling periods.

2.1 Time Delay Model

Most of the nature of time delays has not been completely defined, therefore, it is a possibility to define these as uncertainties for several applications. Since there is not a unique nature of time delays, novel approaches become an affordable and competitive issue on course.

Generally, in common configurations of a NCS (Fig. 2.1), the Round Trip Time (RTT) is defined as the time it takes a package to go from a sensor node to an actuator node. The sensor node sends an information package to the controller node, and then, the controller node computes the control signal and sends it to the actuator node. Thus, the RTT is divided into two kinds of time delays: from the sensor node to controller node (τ_{sc}), and from controller node to actuator node (τ_{ca}).

If the sensor and actuator nodes are the same as case A or all the nodes are completely distributed as case B, the controller node normally is able to measure the τ_{sc}, and estimate τ_{ca}. In case C, if the sensor and controller nodes are the same, $\tau_{sc} = 0$, and τ_{ca} is estimated. Finally, the case D if the controller and actuator nodes are the same, the RTT is simply computed since τ_{sc} can be directly measured and $\tau_{ca} = 0$.

From previous configurations, the first two cases present both time delays τ_{sc} and τ_{ca}, and thus, the case B is the one to be used to apply the methodologies proposed

H. Benítez-Pérez et al., *Control Strategies and Co-Design of Networked Control Systems*, Modeling and Optimization in Science and Technologies 13,
https://doi.org/10.1007/978-3-319-97044-8_2

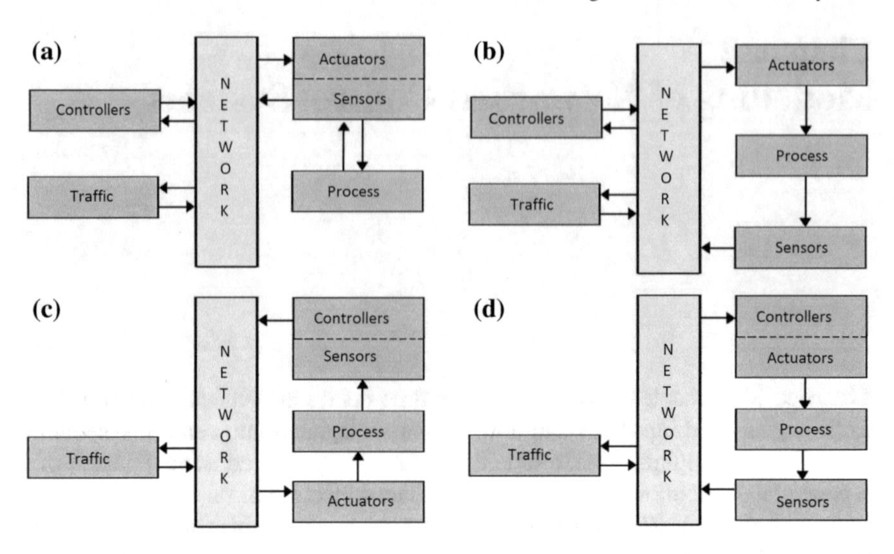

Fig. 2.1 NCS common configurations

in this book. As stated before, only this configuration allows to measure τ_{sc}, but it is necessary to estimate τ_{ca}.

The occurrence of time delays on the nondeterministic networks are generally unknown and time varying and sometimes modelled as a random process with independent distribution. In real NCS implemented over IP networks exhibit time-varying time delays depending on the propagation distance, querying and congestion conditions, etc. This variability may be evaluated and taken into account in the modelling. Figure 2.2 illustrates real RTT delays measured between two x86 PCs interconected with an Ethernet network in the distributed control lab at National Autonomous University of Mexico (UNAM) campus Ciudad Universitaria. It shows the random behaviour in a short time [1].

Figure 2.3 shows a histogram of the RTT time delays from an Ethernet network at UNAM, the histogram is skew to the left indicating a higher probability to have delays shorter than the mean and with a much lower probability to have delays much longer than the mean. This shape of the histogram can be modelled by a random probability distribution.

We use the generalized exponential distribution to describe the RTT delays, this statistical model has the advantage of being simple to compute and easy to implement for real-time prediction. Besides, the parameters can be dynamically adapted, depending on the kind of network and different traffic conditions, significantly reducing the prediction error [2]. These are sufficient reasons to use the generalised exponential distribution as follow:

$$P(\tau) = \begin{cases} \frac{1}{\varphi} e^{-(\tau-\eta)\varphi} & \tau \geq \eta \\ 0 & \tau < \eta \end{cases} \tag{2.1}$$

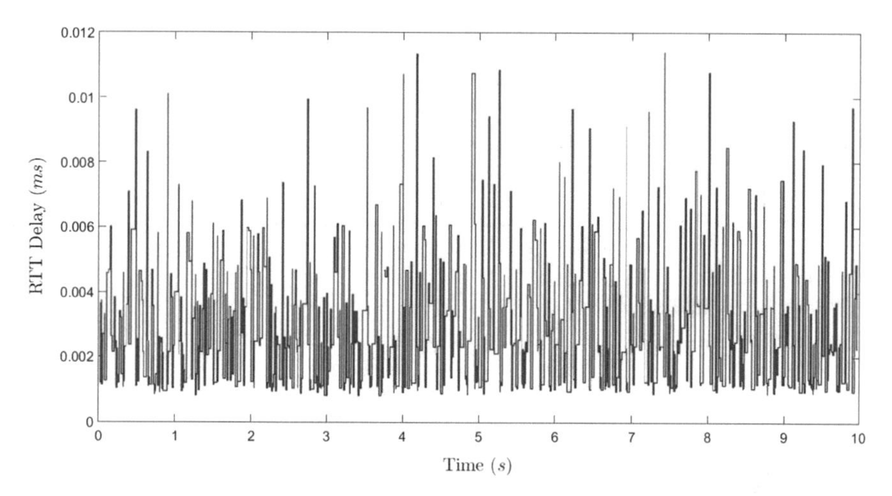

Fig. 2.2 RTT delay measured from an Ethernet network at UNAM

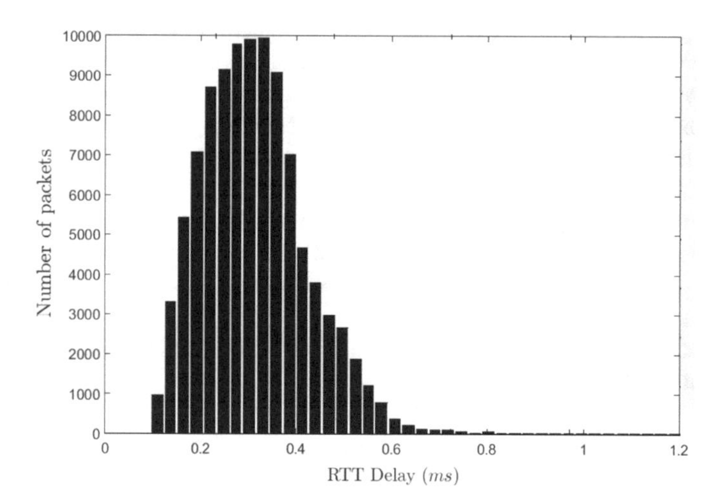

Fig. 2.3 Histogram of RTT delays measured into an Ethernet network at UNAM

where η is a location parameter defining by the median of the time delay and φ is a scale parameter defining by the standard deviation. The sample space is defined by the minimum RTT delay and the maximum allowable time delay bound (MATI) [3].

This distribution is computed for a time window with w previous measurements of time delay $(\tau_{k-w+1}, \ldots, \tau_{k-1}, \tau_k)$, where a expected time delay value for RTT is $E[\tau_{k|w}] = \varphi + \eta$ with variance $Var[\tau_{k|w}] = \varphi^2$ [4].

The design of the model starts by choosing η. The median of the window is a feasible election, since it represents an expected time delay within the window. The distribution is obtained for each one of the values in the time window $(\tau_{k-w+1}, \ldots, \tau_{k-1}, \tau_k)$. $P[\tau_k = \hat{\eta}_{k+1|w}]$ represents the time delay with the maximum probability. Using this value, the expected time delay values of the median $\hat{\eta}_{k+1}$ and standard deviation $\hat{\varphi}_{k+1}$ are computed, and hence:

$$\hat{\eta}_{k+1|w} = w_i | P_{\max_i} [w] \tag{2.2}$$

$$\widehat{\varphi}_{k+1|w} = \sqrt{\text{var}\left\{\tau_{k-w+1}, \ldots, \tau_{k-1}, \tau_k\right\}} \tag{2.3}$$

Being the estimated RTT:

$$\widehat{\tau}_{k+1|w} = \widehat{\varphi}_{k+1|w} + \widehat{\eta}_{k+1|w} \tag{2.4}$$

2.2 Adaptive Fuzzy Model

The strategies for control systems with static time delays is common in literature. Nevertheless, strategies considering variable time delays are uncommon. Among this kind of strategies, it is used optimal control (LQR, LQG) [5, 6], robust control (H_∞) [7, 8], vanishing perturbation [9], and others. These strategies aim to keep a control input resilient to variations of time delays, no matter whether this is an optimal input for the NCS.

In this section, an proposed method is introduced using a fuzzy controller, its advantage is to generate optimal control signals for each time delay value. While, it is robust to variations in such time delays. First, the time delays are analyzed an modeled as Sect. 2.1.

The fuzzy control strategy based on TSK considers a variable structure system representation, this shows as:

$$\mu_{ij}\left(x_i\left(k\right)\right) = e^{-\left(\frac{x_i(k)-\kappa_{ij}}{\sigma_{ij}}\right)^2} \qquad i = 1, \ldots, n,\ j = 1, \ldots, r \tag{2.5}$$

where $x_i(k)$ is the ith state, n is the number of states, r is the number of fuzzy rules, and μ is the related membership function. Thus:

$$g_j = \prod_{i=1}^{n} \mu_{ij}(x_i(k)) \qquad j = 1, \ldots, r \tag{2.6}$$

$$h_j = \frac{g_j}{\sum_{i=1}^{r} g_i} \qquad j = 1 \ldots r \tag{2.7}$$

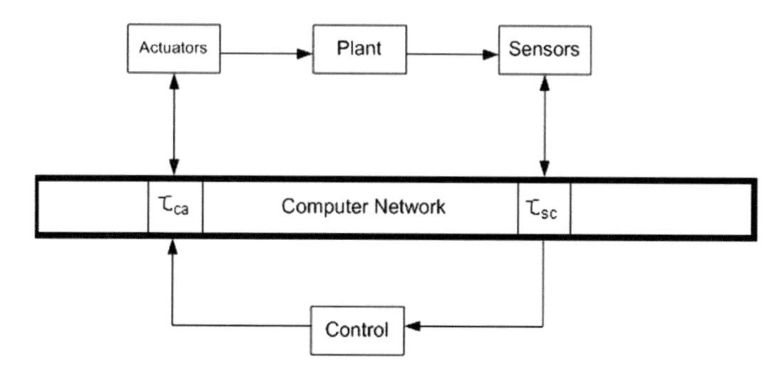

Fig. 2.4 Organization of nodes for a NCS

$$x(k+1) = \sum_{j=1}^{r} h_j (A_j x(k) + B_j u(k)) \tag{2.8}$$
$$y(k) = \sum_{j=1}^{r} h_j C_j x(k)$$

where $x(k) \in \Re^n$ are the system states (fuzzy inputs), n is the number of states, $u(k) \in \Re^m$ are the system inputs, m is the number of system inputs, $y(k) \in \Re^q$ are the system outputs, q is the number of system outputs. Also, $A_j \in \Re^{n \times n}$, $B_j \in \Re^{n \times m}$, and $C_j \in \Re^{q \times n}$ are the plant representation per scenario according to current time delays, where n is the number of states, m is the number of inputs of the system, and r is the number of fuzzy rules, as shown in Fig. 2.4.

Let us define a fuzzy model $x_p(k)$ for the plant. The Eq. 2.9 models some kinds of faults using a vector ρ_j. So, ρ_j^p masks the plant faults and the model (2.9) is considering a local variable structure when faults occur and incorporating the time delays τ^{sc} and τ^{ca}. Remember, the nature of fault is out of the scope of this book. Notice that ρ_j^p is the relation of a fault presence through the event.

$$x_p(k+1) = \sum_{j=1}^{r} h_j (A_j^p x_p(k) + \rho_j^p B_j^p u_c(k - \tau_j^{ca})) \tag{2.9}$$
$$y_p(k) = \sum_{j=1}^{r} h_j C_j^p x_p(k)$$

where the subindex p corresponds to the dynamics of the plant.

From [10], the time delay representation in terms of discrete system in the range $[t_i, t_{i+1}]$ is:

$$B_j^p = \sum_{j=1}^{l_p} \int_{t_i}^{t_{i+1}} e^{\hat{A}_j^p t} \hat{B}_j^p dt \tag{2.10}$$

where $\hat{B}_j^p \in \Re^{n \times m}$ is the input matrix and $\hat{A}_j^p \in \Re^{n \times n}$ is the state matrix of the continuous model of the plant. Remember that l_p is the total number of local time delays that appears per scenario, and they respectively are the source of τ_j^{sc} and τ_j^{ca}. Suppose that the system is in a stable equilibrium point. Thus, the only perturbations are the time delays, and hence, it is possible to assume the outputs (y_p) of the plant.

Since there is no external perturbation, and the dynamics of actuators and sensors are transparent, that is, they do not modify the control and state signals respectively, the outputs are gathered as follows:

$$y_p(k) = \sum_{j=1}^{r} h_j C_j^p x_p(k) \rightarrow u_c\left(k - \tau_j^{sc}\right) = y_p(k - \tau_j^{sc}) \tag{2.11}$$

From this equation, substituting in Eq. 2.9, a complete representation of plant is obtained in terms of a system in equilibrium, given as:

$$x_p(k+1) = \sum_{j=1}^{r} h_j \left(A_j^p x_p(k) + \rho_j^p B_j^p \sum_{i=1}^{r} h_j C_i^p x_c(k - \tau_i^{ca}) \right) \tag{2.12}$$

where x_c is the feedback state for control.

Since the observed states need to be guaranteed, it is necessary to define a group of observers. The proposal of a control law in Sect. 4.2 use a group of observers to guarantee as well the system structure and the local time delays are integrated into the dynamic structure. Therefore, stability, in terms of time delays and misleading structure, should be accomplished. A review of the control law is developed through the following chapters. The related observer states are presented as $z(k)$. In this case N_j, M_j, L_j and T_2 are the observer parameters, to be defined in Sect. 4.2 [11].

2.3 Sampling Frequency Model

One of the strategies implemented for NCS control is feedback scheduling, where the goal is to know the performance of the communication network to modify the transmission of each node and improve the performance of the whole system.

The proposal for feedback scheduling into a real-time distributed system modifies the transmission frequency of each node into a region where the performance in each node is acceptable. This approach is a representation of the transmission frequencies using a time-invariant space state system, whose state variables are the desired transmission frequencies f_d^i with $i = 1, \ldots, n$ of the n nodes involved in the NCS.

The objective of the space state system is to obtain the next desired frequencies. Each node i is bounded into a minimum f_m^i and maximum f_x^i transmission frequency computed offline. The f_m^i are defined by the steady-state stability requirements associated to the node i and the f_x^i are defined by the capacity of the node i itself without generating an excessive transmission cost into the network i.e. in a distributed system with n nodes, each one performing a task t_i with a maximum frequency f_x^i and consumption time per transmission packet ε_i, for $i = 1, 2, \ldots, n$ the following computational constraint needs to be accomplished:

$$U = \sum_{i=1}^{n} \varepsilon_i * f_x^i \leq 1 \tag{2.13}$$

This equation implies that consumptions due to message transmission with the maximum frequencies should not exceed the network utilisation in order to avoid packet losses and time delays.

So, the state variables $x = \{x_1, x_2, \ldots, x_n\}$ are the desired frequencies of the n nodes in the network control system. The objective of the proposed controller is to stabilize the current frequency of the system f_r to the desired reference frequency f_d, through an optimal trajectory using a LQR controller.

By defining a discrete state space model as

$$\begin{aligned} x[k+1] &= Ax[k] + Bu[k] \\ y[k] &= Cx[k] \end{aligned} \tag{2.14}$$

where $A \in \Re^{n \times n}$ is the matrix of relationships between frequencies of the nodes, $B \in \Re^{n \times n}$ is the weighted frequency matrix, $C \in \Re^{n \times n}$ is the output matrix, $x \in \Re^n$ is a real frequency vector, and $y \in \Re^n$ is the next frequency vector.

Let $a_{ij} \in A$ be given by a function of minimal frequencies f_m^i and $b_{ij} \in B$ given by a function of maximal frequencies f_x^i, where $i = 1, 2, \ldots, n$, $j = 1, 2, \ldots, n$ and n nodes:

$$a_{ij} = \varphi(f_m, f_r) \tag{2.15}$$

$$b_{ij} = \gamma(f_m, f_x) \tag{2.16}$$

The elements a_{ij} and b_{ij} in Eq. 2.14 are defined as follows:

$$a_{ij} = \begin{cases} \frac{f_m^i}{f_r^i} & i = j \\ 0 & i \neq j \end{cases} \tag{2.17}$$

$$b_{ij} = \begin{cases} \frac{f_m^i}{f_x^i} & i = j \\ \frac{f_m^i}{f_x^j} & i \neq j \end{cases} \tag{2.18}$$

$$c_{ij} = \begin{cases} 1 & i = j \\ 0 & i \neq j \end{cases} \tag{2.19}$$

This will provide a model to modify the transmission frequency of each node depending on the current frequency. It is a reconfigurable system where indirectly the behaviours of the network imperfections and the NCS system are modified.

2.4 Control-Scheduling Codesign Model

Another approach for controlling NCSs with variable time delays is to take into consideration actuation periods. Here, the objective is to bound the variation space of the time delay to discrete values, making use of an estimated time delay with values that are multiples of the actuation period, and generate optimal control signals for such an estimated time delay. The configuration of a NCS takes into consideration the sensor node driven in time, whereas the controller and actuator nodes are event-driven. Using a space-state model of a discrete time linear system in invariant discrete time, with a sampling period h [12]:

$$\begin{aligned} x(k+1) &= \Phi(h)x(k) + \Gamma(h)u(k) \\ y(k) &= Cx(k) \end{aligned} \tag{2.20}$$

where $x(k)$ is the state of the plant, $u(k)$ and $y(k)$ are respectively the inputs and outputs of the plant, matrix $C \in \Re^{p\times n}$ is the output matrix, and matrices $\Phi(h)$ y $\Gamma(h)$ are obtained using:

$$\Phi(h) = e^{Ah} \qquad \Gamma(h) = \int_0^h e^{A} B ds \tag{2.21}$$

with $A \in \Re^{n\times n}$ and $B \in \Re^{n\times m}$ as the matrices of the system and input of the model in continuous form:

$$\begin{aligned} \dot{x}(t) &= Ax(t) + Bu(t) \\ y(t) &= Cx(t) \end{aligned} \tag{2.22}$$

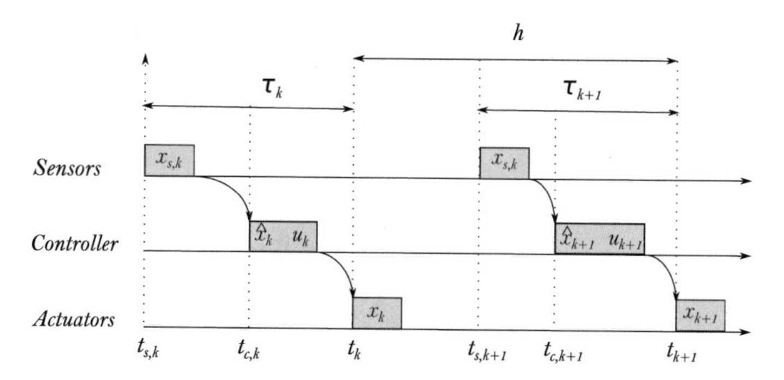

Fig. 2.5 Execution time of control tasks with instants of actuation

In the specific case of NCS, I/O latencies result from computing the control algorithm or due to the insertion of the network in the control loop. The standard model that comprises the time delay τ, with $\tau < h$, is [12]:

$$x(k+1) = \Phi(h)x(k) + \Phi(h-\tau)\Gamma(\tau)u(k-1) + \Gamma(h-\tau)u(k) \tag{2.23}$$

The model in Eq. 2.23 is the simplest form used for analyzing and designing controllers for NCS. This model assumes a reference time, given by the sampling instants with a fixed time delay, starting at sampling, and until actuation. But if the time delay is variant and larger than the sampling period and/or the sampling interval is variant, then this model is not very feasible for NCS [13].

Here, a execution time of control tasks that synchronizes the operation of each control loop at the instant of actuation is proposed (Fig. 2.5). The sampling instants are labelled by $t(s,k)$ where the system states $x(s,k) \in [t(k-1), t(k)]$ are obtained, and the instants of actuation are labelled by $t(k)$ where the control signals $u(k)$ are applied. Hence, the time elapsed between consecutive instants of actuation $[t(k-1), t(k)]$, has a actuating period h. The difference between the sampling instant $t(s,k)$ and the following actuation instant $t(k)$ is a variable time delay:

$$\tau(k) = t(k) - t(s,k) \tag{2.24}$$

and it is used for estimating the state in the actuation instant:

$$\hat{x}(k) = \Phi(\tau(k))x(s,k) + \Gamma(\tau(k))u(k-1) \tag{2.25}$$

A control strategy relies on the time reference given by the instant of actuation. In addition, samples are not required to be strictly periodic, because $\tau(k)$ in Eq. 2.24 can vary at each closed-loop operation. The interested reader is referred to [14] for further reading on this control model.

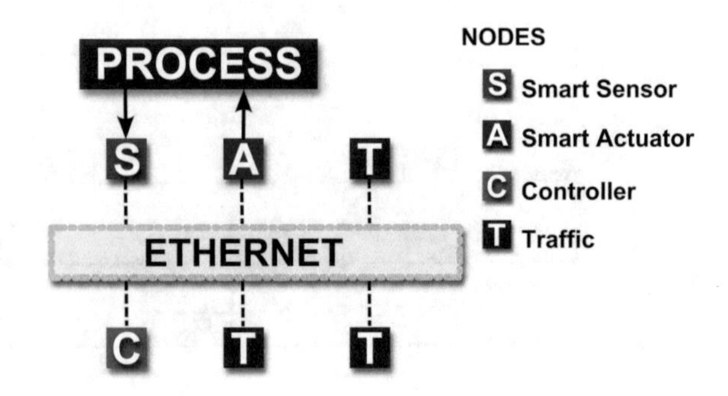

Fig. 2.6 NCS configuration

Each instant of actuation, the actuator node generates the next actuation instant $t(k)$, which is sent to the sensor after applying the control signal $u(k-1)$ to the plant. Upon reception of this message or next sampling period, the sensor samples the plant $x(s,k)$, and records the absolute sampling time $t(s,k)$. The latter, together with $t(k)$, is used to compute $\tau(k)$. Both $x(s,k)$ and $\tau(k)$ are sent to the controller node. Upon reception of this message, the controller node estimates the plant state that applies at $\tau(k)$ (Eq. 2.25) with the next estimated time delay $\tau(k+1)$, and computes a control command $u(k)$ designed in Chap. 4. This is sent to the actuator that applies it to the plant at the synchronised actuation instant.

The NCS architecture (Fig. 2.6) has four types of nodes connected to an Ethernet network: sensor, controller, actuator, and traffic nodes. A control loop comprises communication between sensor-controller and controller-actuator nodes. Traffic nodes send periodic or sporadic packets into network, like for example, other control loops or monitoring.

All nodes send User Datagram Protocol (UDP) packets to avoid double traffic into network. However, it is not possible to know whether there is packets loss or a maximum time delay since there is no acknowledgement that the sent packet has been received.

Here, network-induced time delays and variable sampling intervals smaller than a sampling period are compensated by the one-shot model, and the fuzzy model compensates time delays longer than such a sampling period.

The time delay is estimated in the controller using Eqs. (2.1–2.4) (Fig. 2.7). When the control node receives a new packet, this contains the state $x(s,k)$ and the time delay $\tau(k-1)$. Once the time delay $\tau(k-1)$ is received, an exponential distribution algorithm [15] is used to estimate the next time delay $\hat{\tau}(k)$. This algorithm is widely used to estimate delays in real time, where an offline statistical analysis characterises mean and standard deviation $q_E(\eta,\phi)$ of time delays data τ with multiple scenarios of traffic. Those are used to form a generalized exponential distribution with a probability density function (Eq. 2.1).

Once the next time delay $\hat{\tau}(k)$ of the system is estimated, it is used to generate a control signal. This controller starts with a fuzzy model of the NCS, designed as follows.

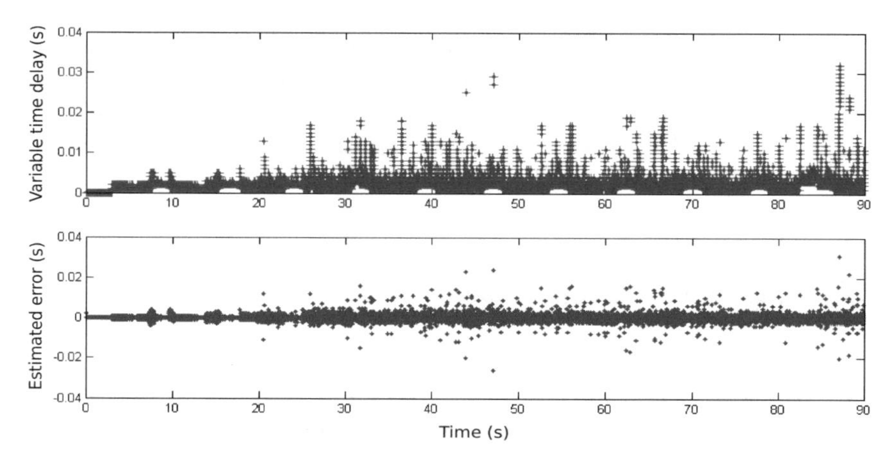

Fig. 2.7 Variable time delay (τ_k) behaviour and estimation error with algorithm in Sect. 2.1

The fuzzy model (Eq. 2.25) is TSK type [16], with the time delay $\hat{\tau}(k)$ as input of the antecedent part, and linear discrete models with different sampling periods h_j as consequent part. Thus, defining r fuzzy rules, the jth rule has the form:

$$if\ \hat{\tau}(k)\ is\ \alpha_j(\hat{\tau})\ then\ x_j = \Phi(h_j)x(s,k) + \Gamma(h) \tag{2.26}$$

where $x(k) \in \Re^n$ is the state vector of system, $u(k) \in \Re^n$ is the input vector of process, $\alpha_j(\hat{\tau})$ is the jth membership function of the estimated time delay $\hat{\tau}(k)$.

The overall fuzzy model is:

$$\hat{x}(k + \hat{\tau}(k)) = \sum_{j=1}^{r} \Psi_j(\Phi_j x(s,k) + \Gamma_j u(k)) \tag{2.27}$$

where the normalized fire strength Ψ_j is:

$$\Psi_j = \frac{\alpha_j}{\sum_{s=1}^{r} \alpha_s} \qquad \Psi_j \geq 0 \qquad \sum_{j=1}^{r} \Psi_j = 1 \tag{2.28}$$

and $\alpha_j = exp(-(\hat{\tau}(k) - \rho_j)^2/\sigma_j^2)$ is a Gaussian membership function with parameters (ρ_j, σ_j). (Φ_j, Γ_j) are the matrices of jth linear discrete model discretised with a sampling period h_j $(j = 1, \dots, r)$. The discrete local models are:

$$x_j = e^{Ah_j}x(k) + \int_0^{h_j} e^{As} ds Bu(k) = \Phi_j x(k) + \Gamma_j u(k) \tag{2.29}$$

so, (h_j, ρ_j, σ_j) for $j = 1, \dots, r$ are assigned by user according to offline time delay measurement.

With this fuzzy model, the estimated state of a system is obtained by compensating the time delays and variable sampling intervals. The action is to smoothly switch between discrete models, to generate the best estimate of state, according to the estimated time delay $\hat{\tau}(k)$. Once designed the fuzzy model (2.26) using the estimated time delay $\hat{\tau}(k)$, a fuzzy controller is proposed in Chap. 4.

2.5 Neuro-Fuzzy Identification

The objective of the neuro-fuzzy model is to define the best linear model for each sampling period representing the system. This model estimates the system states according to the possible time delay calculated by an exponential distribution function defined in Sect. 2.1.

The neuro fuzzy model (Eq. 2.30) is TSK type [17] with variable sampling period $\hat{\tau}(k)$ as antecedent input, and discrete models with different sampling periods T as consequent part. So, defining r fuzzy rules, the jth rule has the form:

$$\text{if } \hat{\tau}(k) \text{ is } \alpha^j(\hat{\tau}) \text{ then } \mathrm{x}^j\left(k + T^j\right) = \Phi^j x(k) + \Gamma^j u(k) \tag{2.30}$$

where $x(k) \in \Re^n$ is the state vector of the system, $u(k) \in \Re^m$ is the input vector of the plant, α^j is the jth membership function of variable sampling period $\hat{\tau}(k)$.

The overall fuzzy model is:

$$\hat{x}\left(k + \hat{\tau}\right) = \sum_{j=1}^{r} \psi^j \left(\Phi^j x\left(k\right) + \Gamma^j u\left(k\right)\right) \tag{2.31}$$

where the normalized fire strength ψ^j is:

$$\psi^j = \frac{a^j}{\sum_{s=1}^{r} a^s}, \quad \psi^j \geq 0, \ \sum_{j=1}^{r} \psi^j = 1, \ a^j = \exp\left(-\frac{\left(\hat{\tau}_k - \rho^j\right)^2}{\left(\sigma^j\right)^2}\right) \tag{2.32}$$

where α^j is a Gaussian membership function with parameters (ρ^j, σ^j), and $\Phi^j \in \Re^{n\times n}$ and $\Gamma^j \in \Re^{n\times n}$ are the matrices of jth linear model, with m outputs and periods $T = \{\rho^1, \rho^2, \ldots, \rho^r\}$.

The system states are approximated with this fuzzy model. The objective is to smoothly switch between discrete models, to generate the best estimate of states according to the variable sampling period$\hat{\tau}_k$. However, there are some parameters such as $(\Phi^j, \Gamma^j, \rho^j, \sigma^j)$ that need to be tuned using a neuro-fuzzy approach [18, 19]. This section uses data of the simulated system with some specific time delays to get the model parameters, although, alternatively techniques like LMI may be used and reviewed in other works as [20]. The neuro-fuzzy model (Eq. 2.30) is identified using inputs-states data with measured variable sampling period $D = \{x_l, u_l, \tau_l | l = 1, \ldots, L\}$, obtained from a Truetime simulation of the system [21], with a PID controller and a traffic node on an Ethernet network.

The identification procedure has two algorithms: a clustering algorithm, to create new rules, and a training algorithm, to update the model's parameters. The procedure is repeated ϕ-epochs. M is maximum firing strength of all rules for variable sampling period and K_d is its threshold. K_e is a threshold of maximum model error. J is the performance index to update the matrices (Φ^j, Γ^j).

Procedure: The first fuzzy rule is created with the discrete model, with sampling period ρ_0, which is the variable sampling period mean, and σ_0 is the predetermined width. So, matrices Φ^1 and Γ^1 of the first rule are:

$$\Phi^1 = e^{(A\rho_0)} \tag{2.33}$$

$$\Gamma^1 = \int_0^{\rho_0} e^{As} ds B \tag{2.34}$$

$$\rho^1 = \rho_0 = mean\,(\tau) \qquad \sigma^1 = \sigma_0 \tag{2.35}$$

A new rule is created using the ε-completeness criterion, which states that, for any input within operation range, there is a rule with fire strength α^j greater than a threshold K_d [22]. So, if maximum firing strength M of all rules is less than K_d then a new rule is created.

$$M = \max_j \left(\alpha^j\right) \tag{2.36}$$

$$M < K_d \tag{2.37}$$

With this clustering algorithm, it is ensured that all variable sampling periods of the training data are represented by the antecedent part of the neuro-fuzzy model.

In the training algorithm, the parameters of local models are estimated using error criterion. It says that, if an output error e is less than a threshold K_e (Eq. 2.40), the parameters should be adjusted. The threshold K_e is decreased according to the current epoch f.

$$e = \left|y\,(k+1) - \hat{y}\,(k+1)\right| \tag{2.38}$$

$$e > K_e \tag{2.39}$$

$$K_e = \frac{e_{\max} + e_{\min}}{\phi - 1}(f - 1) + e_{\max} \tag{2.40}$$

where e_{max} is the final error expected at the end of training, e_{min} is the initial error obtained in the first epoch without training, f is the current training epoch, and ϕ is the total of epochs.

The backpropagation approach is used for the adjustment of the parameters. So, a performance index J is used to adjust the model with objective to minimise J (modelled error).

$$J = \frac{1}{2}\left\|\hat{x} - x\right\|^2 = \frac{\sum_{p=1}^{n}\left(\hat{x}_p - x_p\right)}{2} \tag{2.41}$$

The coefficients $a^j_{p,q}$ and $b^j_{p,o}$ of the system Φ^j and control Γ^j matrices, respectively, as well as the centres and standard deviations are adjusted by:

$$a^j_{q,p}(k+1) = a^j_{q,p}(k) - \eta_a\left(\hat{x}_p(k+1) - \hat{x}_p(k+1)\right)\psi^j x_q(k) \tag{2.42}$$

$$b^j_{p,i}(k+1) = b^j_{p,i}(k) - \eta_b\left(\hat{x}_p(k+1) - \hat{x}_p(k+1)\right)\psi^j u_i(k) \tag{2.43}$$

$$\begin{aligned}\rho^j(k+1) = \rho^j(k) - \eta_\rho\left(\hat{x}(k+1) - \hat{x}(k+1)\right)& \\ \left(\Phi^j\hat{x}(k) + \Gamma^j u(k)\right)\left(\psi^j - \left(\psi^j\right)^2\right)\left(-2\frac{\hat{\tau} - \rho^j}{\left(\sigma^j\right)^2}\right)&\end{aligned} \tag{2.44}$$

$$\begin{aligned}\sigma^j(k+1) = \sigma^j(k) - \eta_\sigma\left(\hat{x}(k+1) - \hat{x}(k+1)\right)& \\ \left(\Phi^j\hat{x}(k) + \Gamma^j u(k)\right)\left(\psi^j - \left(\psi^j\right)^2\right)\left(-2\frac{\left(\hat{\tau} - \rho^j\right)^2}{\left(\sigma^j\right)^3}\right)&\end{aligned} \tag{2.45}$$

where η is the learning rate for each parameter, $q, p = 1, \ldots, n$, $o = 1, \ldots, m$; and $j = 1, \ldots, r$. The data used in here is the result of analysis of general scenarios considering local time delays and ethernet channel communication.

Once designed and identified, the fuzzy model (Eqs. 2.30, 2.31, 2.32) is presented a fuzzy controller in Chap. 4, using estimated variable sampling period $\hat{\tau}(k)$.

2.6 Concluding Remarks

This chapter presents a statistical model of the time delay through an exponential probability density function that allows estimating the time delay for compensation purposes, the model assumes a variable time delay and greater than the sampling period of the system, Model is validated using experimental data. Three design methodologies are also presented.

The first method presents a model using fuzzy theory to incorporate the estimated time delay with several space-state models. This allows improving the state estimation to be used in the controller with minimal computational cost.

The second method presents a sampling frequency model through a state-space model that represents the dynamic behaviour of the transmission frequency as a function of the actual transmission frequency into the network with longer time delays or lost deadlines.

The third method presents a methodology of codesign where through two fuzzy models are trade-off the control and scheduling performance. The first model incorporates linear models with different sampling periods obtaining delay-dependent system states. The second model uses QsC and QsS to schedule the next stable actuation period and minimise the system error and the effect of network uncertainties.

Finally, a neural strategy is presented to obtain the fuzzy rules that allow to identify the complete NCS system and to improve the estimation of the states of the system.

References

1. Carnevale, D., Teel, A.R., Nesic, D.: A Lyapunov proof of improved maximum allowable transfer interval for networked control systems. IEEE Trans. Autom. Control **55**, 892–897 (2007)
2. Mendez-Monroy, E.: Codiseño de Sistemas de Control en Red Compensando Imperfecciones Acotadas de Tiempo Inducidos por la Red. Posgrado en Ingenieria, Campo de Conocimiento de Electrica, PhD, UNAM. 22 Junio (2012)
3. Ma, C., Chen, S., Liu, W.: Maximum allowable delay bound of networked control systems with multi-step delay. Simul. Model. Pract. Theor. **15**(5), 513–520 (2007)
4. Lian, F., Moyne, J., Otanez, P., Tilbury, D., Moyne, J.: Design of sampling and transmission rates for achieving control and communication performance in networked multi-agent system. In: Proceedings of American Control Conference, pp. 3329–3334 Denver, USA (2003)
5. Gupta, V., Spanos, D., Hassibi, B., Murria, M.R.: On LQG control across a stochastic packet-dropping link. In: Proceedings 2005 American Control Conference, pp. 360–365, Portland, USA (2005)
6. Cloosterman, M., van de Wouw, N., Heemels, W., Nijmeijer, H.: Stabilization of networked control systems with large delays and packet dropouts. In: American Control Conference, pp. 4991–4996 (2008)
7. Meng, X., Lam, J., Gao, H.: Network-based H∞ control for stochastic systems. Int. J. Robust Nonlinear Control **19**(3), 295–312 (2009)
8. Jiang, X., Han, Q.L., Liu, S., Xue, A.: A new H∞ stabilization criterion for networked control systems. IEEE Trans. Autom. Control **53**(4), 1025–1032 (2008)

9. Millan P., Orihuela L., Vivas c., Rubio F.R.: Control Optimo-L2 Basado en Red Mediante Funcionales de Lyapunov-Krasovskii, Revista Iberoamericana de Automatica e Informatica Industrial RIAI **9**(1), pp. 14–23 (2012)
10. Ogata, K.: Discrete-Time Control Systems. Prentice-Hall Inc., Upper Saddle River, NJ, USA (1987)
11. Esquivel-Flores, O.A.: Estudio de Sistemas Multi-agentes Reconfigurables. Posgrado en Ciencias e Ingenieria de la Computacion, UNAM. 23 Enero (2013)
12. Aström, K.J., Wittenmark, B.: Computer-Controlled Systems: Theory and Design, 2nd edn. Prentice-Hall Inc., Englewood Cliffs, NJ (1990)
13. Nilsson, J.: Real-Time control systems with delays; PhD Thesis, Dept. Automatic Control. Lund Institute of Technology, Lund, Sweden (1998)
14. Marti, P., Velasco, M.: Toward flexible scheduling of real-time control tasks: reviewing basic control models. In: Proceedings of the 10th International Conference on Hybrid Systems, Computation and Control. LNCS (2007)
15. Tipsuwan, Y., Chow, M.Y.: On the gain scheduling for networked pi controller over IP network. IEEE/ASME Trans. Mech., **9–3** (2004)
16. Tanaka, K., Wang, H. O.: Fuzzy Control Systems Design and Analysis: A Linear Matrix Inequality Approach. Wiley (2001)
17. Mendez-Monroy, P.E., Benitez-Perez, H.: Fuzzy control with estimated variable sampling period for non-linear networked control systems: 2-DOF helicopter as case study. Trans. Inst. Meas. Control **34**(7), 802–814 (2012)
18. Mendez-Monroy, P.E., Benitez-Perez, H.: Identification and control for discrete dynamics systems using space state recurrent fuzzy neural networks. In: Electronics, Robotics and Automotive Mechanics Conference (CERMA 2007), pp. 112–117. IEEE (2007)
19. Gonzalez-Olvera, M.A., Tang, Y.: A new recurrent neurofuzzy network for identification of dynamic systems. Fuzzy Sets Syst. **158**(10), 1023–1035 (2007)
20. Benitez-Perez, H., Benitez-Perez, A., Ortega-Arjona, J.: Networked control systems design considering scheduling restrictions and local faults using local state estimation. IJICIC **9–8**, 3225–3239 (2013)
21. Cervin, A., Henriksson, D., Lincoln, B., Eker, J., Arzen, K.-E.: How does control timing affect performance? In: Analysis and Simulation of Timing using Jitterbug and Truetime. Control Systems, IEEE **23**(3), 16–30 (2003)
22. We, S., Er, M.J., Gao, Y.: A fast approach for automatic generation of fuzzy rules by generalized dynamic fuzzy neural networks. IEEE Trans. Fuzzy Syst. **4**(4), 578–598 (2001)

Chapter 3
Distributed Systems Modelling

Abstract Distributed systems have a lot of applications, commonly with multitasks where the information exchange is high through a communication network, this exchange presents inherent time delays that degrade the system. For obtaining an acceptable performance in the distributed system, time delays need to be taken into consideration during design. This chapter is devoted to reviewing some representations of time delays in function of some features as the operative system, network, scheduling. These include the scalability, concurrency, feasibility and extensibility. Several situations are reviewed, such as aperiodic communications or consensus needs, among other situations. Finally, an introduction to a helicopter dynamic model is given, focusing on real-time representation with a relationship matrix of transmission frequencies through True-time simulation.

3.1 Classical Distributed Systems

Distributed systems are composed of autonomous elements with the capacity of processing, storage, and communicating among themselves; normally, decisions are taken following consensus strategies, as well as optimising between scales, from local to global. Taking this definition as valid, a distributed system, thus, can be modelled following the relations between elements in terms of simple interactions or described as independent components pursuing a common goal, that may be known by all of them. An important feature is the communication media between elements, which can be wire-based or wireless. In either case, these conditions represent dependencies through the dynamics of the distributed system.

Distributed systems can be represented as Markov chains or Bernoulli processes. As such, they are feasible depending on those structures and meaningful to highlight features such as time delays or known global data losses, without the compromise of certainty on system response [1, 2]. Other representations like classical real-time systems, provide the advantage of certainty over several system conditions, and time response. The interaction amongst computing elements in a distributed system directly depends on the communication channels, as well as the processing capabilities per local element. In this case, complexity becomes an issue, in terms of

H. Benítez-Pérez et al., *Control Strategies and Co-Design of Networked Control Systems*, Modeling and Optimization in Science and Technologies 13,
https://doi.org/10.1007/978-3-319-97044-8_3

modelling at different scales. One of the challenges is to achieve synchronisation, in order to accomplish sharing resources, like memory use.

A major motivation for the use of distributed systems has been sharing resources, but keeping services reliable. For this, some techniques may be used, such as replication, in the case that a component of the distributed system fails. When designing a distributed system, certain features have to be taken into consideration so applications have a reliable performance. This has to do with aspects such as diversity of computing equipment, communication technology, different operating systems, application availability, limitations of information flow, and safe transmission mechanisms. The field of Distributed Systems is in charge of studying and proposing solutions to the problems present in the implementation of distributed applications. Such applications have features that depend on the systems architecture, and by themselves, they are a real challenge for implementation. These features, in general, are [3]:

- *Scalability*. A distributed system is scalable if it remains effective and efficient even if there is an increase or decrease in the number of resources or users.
- *Security*. A distributed system is secure if it counts with mechanisms that ensure data confidentiality, integrity, and availability.
- *Extensibility*. A distributed system is extensible if there are several ways to make it grow or re-implemented. The extensibility of a distributed system is determined by the number of new resources that can be aggregated and available to be used by different clients.
- *Fault management*. A distributed system is fault-tolerant while it has mechanisms that prevent or correct errors generated within the system. It is vital that distributed systems isolate a fault since its components are physically separated; techniques such as replication and fault masking are suitable for this systems.
- *Concurrency*. A distributed system is naturally concurrent since it is composed of components and resources that simultaneously execute. Resources can be shared by different components, so it may be the case that several components attempt to access a single shared resource at the same time. Hence, it is necessary that such a resource guarantees the correct function for each access. Several techniques are used to guarantee data consistency, by applying synchronisation mechanisms such as semaphores, flags, and message passing.
- *Transparency*. A distributed system is transparent if it covers the separation of its components from the point of view of its users. The distributed system is seen as a whole, not as a set of independent components.
- *Heterogeneity*. A distributed system is composed of autonomous computing resources, which in general are diverse. The diversity may depend on the kind of network, operating system, programming language, and implementation of its applications.
- *Quality of Service (QoS)*. Users benefit from the services provided by the Distributed System, and its quality. The main properties of the systems that affect the quality of service are reliability, security, and performance. Adaptability to satisfy the changing configurations of the system, as well as resource availability, are both important aspects of quality of service. The performance related to the quality of

service is originally defined in terms of response and computational throughput. However, it has been redefined in terms of capacity for achieving time guarantees.

In general, there are three types of distributed systems: distributed computing systems, distributed information systems, and embedded distributed systems [4].

- The distributed computing systems are subdivided into cluster computing and grid computing. Cluster computing is a collection of workstations or personal computers with similar features, connected through a high-speed local network, with each node executing the same operating system. In grid computing, systems are built like federations of computer systems, in which each system is ruled by a different administration, and they can be totally different from each other in terms of the hardware, software, and network technology.
- Distributed information systems are those in which there is a large number of network applications, and there is a need to achieve interoperability between different components. An example of these systems is a Database application.
- Embedded systems are also known as pervasive systems. These systems are composed of small, mobile, battery-operated devices, which make use of wireless communication (although not all these features have to be present in the same device). Examples of these devices are PDAs, smartphones, tablets, health-care devices connected through a body area network, which monitors several organs [4], and sensor networks, used for information processing.

Descriptive models of distributed systems. Reference [3] provide three descriptive models under which the design of a distributed system can be characterised and analysed:

1. Physical model. Encompasses the types of computers, devices, and the interconnectivity that constitute a distributed system.
2. Architectural model. Describes a system in terms of computational tasks and the communication that individual components or groups of components perform.
3. Fundamental model. Consists of an abstraction used to describe solutions to particular problems faced by the distributed systems.

This book studies some features of the architecture and the process of interaction in distributed systems. As follows, several aspects of architectural models are reviewed, aim for defining certain elements on which the analysis and design of reconfiguration of distributed systems are based. The architecture of a distributed system is considered as the structure of the system in terms of its individual components and their interactions. Reference [3] propose four elements that constitute the modern distributed architectures: the kind of entities that communicate through a distributed system, how these entities are communicating (communication paradigms), the specific task that each component performs, and the mapping of tasks on each component within the system.

- Communication entities. From a system point of view, the communicating entities in a distributed system are processes with communication paradigms between them. From a programming point of view, these are objects, components, and nowadays, network services.
- Communication paradigms. Three types of communication paradigms are considered: interprocess communication, remote invocation, and indirect communication. Interprocess communication refers to a low-level communication between the processes of the distributed system. It includes message passing primitives, direct access to APIs, and multicast. Remote invocation is the most common communication paradigm in distributed systems. It covers a wide range of techniques, based on the bidirectional information exchange between entities, and gives place to the operation, procedure, or remote method. These two previous communication paradigms constitute the exchange in a point-to-point, bidirectional way: sender-receiver or client-server. In contrast, indirect communication includes one-to-many, many-to-one, and many-to-many communications, such as group communications, message queues, and distributed shared memory, among others.
- Roles and functions. In a distributed system, processes, components, or services interact with each other to perform several activities; doing this so, they adopt two architectural styles derived from the individual role: client-server and peer-to-peer.
- Mapping. This takes into consideration the place in which objects or services are allocated within the physical infrastructure of the distributed system. Such a decision directly impacts on the complexity of the network that connects a group of computers. Mapping is crucial for performance, reliability, and security of the distributed system. The mapping strategies are focussed on mapping services on multiple servers, storage of objects in cache, use of mobile code, and mobile agents.

3.2 Classical Description

A common description of distributed systems is that in which several nodes gather around a common communication media used for sending and receiving information in common terms. Here, wire-based networks provide us with a wide use of channel management, in order to carry out several communication tasks. Time delays appear, then, as a result of interactions between communication channels, as well as the effect of local concurrency (see Fig. 1.4). The time required to complete a control cycle with feedback is given as:

$$T = nt_s + nt_{cm}^{sc} + t_c + t_{cm}^{ca} + t_a \qquad (3.1)$$

where:

- t_s is the time taken by the sensors,
- n is the number of sensors in the system,
- t_{cm}^{sc} is the communication time between sensor and controller,

- t_c is the time to calculate the control law,
- t_{cm}^{ca} is the communication time between controller and actuator, and
- t_a is the time taken by the actuator.

Notice that the n sensors of the system are sequentially read since there is a small drift between the startup of each sensor. This impacts as n sense times are required, proportional to such a drift if fault scenario is present.

When failure masking is included, the time needed to complete a control cycle with feedback is:

$$T = nt_s + nt_{cm}^{sc} + t_{cm}^{fst} + t_{cm}^{fsc} + t_c + t_{cm}^{ca} + t_a \quad (3.2)$$

where t_{cm}^{fst} is the communication time between sensors and the masking unit, and t_{cm}^{fsc} is the time between the occurrence of a failure (which is out of the scope of this book) and consensus achievement. The reconfiguration process consists of modifying the periods and consumptions of tasks, to counteract the effect of any event which may affect the scheduling condition. Such a modification requires a real-time calculation of new parameters for the tasks. The scheduling of tasks is analysed by obtaining a feasible set of periods, which assure restriction [5] regarding a scheduling algorithm with fixed priorities, such as Rate Monotonic [6]. Obtaining this set of periods depends on an additional set of periods, which serve as a maximum bound that can be taken in a different way depending on the case study.

One of the main contributions of this book consists of showing that reconfiguration aids with diminishing the non-linear effects of the system when there is a considerable number of time delays. An adjustment of the sampling periods results in a smaller time delay in the system's performance, that is relevant to expose; in the same way, an additional time delay is induced, due to the reconfiguration:

$$T_r = t_{cp}^r + t_{rp}^r \quad (3.3)$$

where t_{cp}^r is the calculation time of the new periods, and t_{rp}^r is the time needed for changing the periods.

Local scheduling is not performed through consensus among the agents. However, it is necessary that consensus is assured, to achieve a positive effect in the reconfiguration. This kind of reconfiguration, then, is obtained through consensus, as presented in Chap. 5.

Interaction. Distributed systems are composed of multiple processes, and their interactions are complex. The state and behaviour of the processes can be described using a distributed algorithm. This algorithm defines the steps within each process and the message transmissions between processes for coordinating their activity. The transmission rate of each process and the transmission time is not deterministic, and thus, it is difficult to describe the states of a distributed algorithm. Hence, it is necessary to manage the failures of one or many processes, or the failures in the transmissions.

The interaction between processes performs all the activity within a distributed system. Two factors affect the process interaction within a distributed system: the communication and the notion of global time.

The communication in a computer network has performance features regarding latency, bandwidth, and jitter [7]:

1. The time delay from the beginning of a message transmission in a process and the beginning of its reception by another process is known as latency. Latency includes:
    a. The time to transmit the first bit of the message through the network until it reaches its destination.
    b. The time delay due to accessing the network, which considerably increases when there is a high traffic load.
    c. The time required for the communication services of the distributed system for sending and receiving messages.
2. The total amount of information that the network is able to transmit in a given time frame is known as bandwidth.
3. Jitter is the variation in time when delivering a series of messages. Its value tends to be smaller than the sampling time.

Each node in a distributed system has its own internal clock, which can be used by two local processes to obtain a current time value. Nevertheless, two processes in different nodes do not have access to a common clock. Thus, two processes in different nodes can only associate time-stamps for their events. However, if these two processes read their local clocks at the very same time, they may have a different time value. This is due to the clock drift, which refers to the rate with which a computer clock deviates from an "exact" reference clock. Hence, in a distributed system it is difficult to establish limits to the time that may take the execution of processes, the delivery of a message, or the clock drift. Therefore, to manage the time within a distributed system, two submodels are commonly used [3, 8]:

1. A synchronous model is used when the execution time due to the action of its processes has known superior and inferior bounds. Thus, each message sent through the network is received in a known bounded time, and each process has a local clock whose clock drift rate regarding the reference time has a known bound. The immediate problem with this model is to guarantee the bounds.
2. An asynchronous model has no bounds for the execution speed of processes. Each action may take an arbitrary time. There is a time delay in the transmission of messages, which can be arbitrarily long, and the clock drift is variable. This model does not suppose anything about the time involved in the executions.

Nowadays, distributed systems are asynchronous due to the need of processes and communication channels to share the same network. However, several design problems cannot be solved using an asynchronous approach, in particular when some aspects of time are involved. Nevertheless, it is important to highlight that in reality, distributed systems are hybrid, this is, synchronous in their internal processes, while asynchronous during communication with external processes.

Because clocks cannot be perfectly synchronised in a distributed system, Lamport [9] proposes a logic clock model which can work to provide order between events of processes on different nodes within a distributed system.

Although faults events are out of the scope of this book, a review of fault situations for distributed systems is given, to explore sporadic situations. In a distributed system, channels of communication tend to fail, this is, they do not behave as correctly or desired. The fault model described by Coulouris et al. [3] define the way in which a fault can occur, aiming to understand the side effects of such a fault. Hanzilacos and Toueg [8] provide a classification of faults in processes and communication, retaken by [3], presented as omission faults, arbitrary faults, and time faults.

Omission faults appear when the activity in the processes or communication channels is not carried out. The most frequent fault in processes is crashing. This means that the processes are halted, and it does not perform any action. The detection method for a crash is based on the use of timeouts. In such a method, the process is allowed for a fixed period of time, waiting for something to happen. In asynchronous systems, a timeout may only indicate that a process does not respond. This may be caused by a crash, a slow execution, or a message that has not arrived.

The design and implementation of a NCS require an appropriate integration of the control system and the communication network. Particularly, selecting an appropriate network for control is essential for designing the NCS. It is necessary to understand the features presented in each communication network, as well as its scheduling protocols which define properties such as packet priority, time delay characteristics, packet loss, etc. The time parameters that influence control applications are affected by the data transmission rate, the packet size, and the communication protocol. Compared with data networks, control networks should cope with some features, such as:

1. Fixed and strict sampling periods. In their design, most control loops are supposed to use constant sampling periods for stability analysis. In general, real sampling periods are short and, in addition to the great number of control loops, require a high-speed transmission rate.
2. Short packet length. The length of a packet is determined by the kind of network and the protocol used and varies from a few bytes to some thousand bytes. For control systems, the size of the packets tends to be relatively short.
3. Critical Real-time requirements. Some requirements, such as reliability, latency, fault-resilient, etc., are more critical in a NCS than in data networks.

Control networks that employ feedback are generally based on the following network communication protocols: Ethernet (IEEE 802.3), Token Bus (IEEE 802.4), Token Ring (IEEE 802.5), and DeviceNet (CAN—Controller Area Network) (ISO 11898). The main features of these control networks are summarised as follows and affect the performance of a NCS. At the end of this section, there is a further detailed comparison between the Ethernet, Token Bus, and CAN protocols.

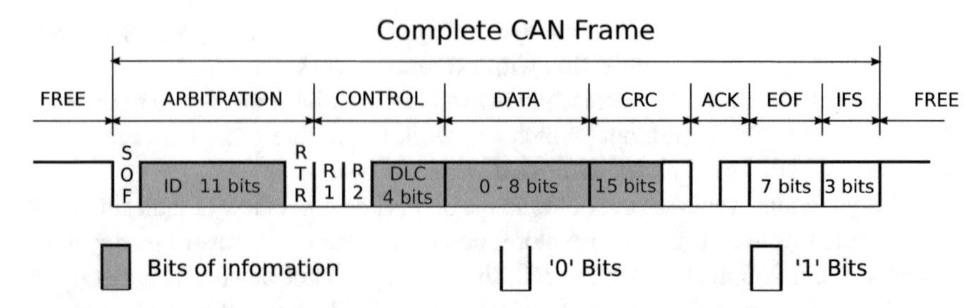

Fig. 3.1 Basic frame format for the CAN standard

DeviceNet. DeviceNet is a protocol in the application layer with a broad use in manufacturing applications. Its specifications are based on the serial communications standard CAN [6, 10], mainly developed for applications in the automotive industry, and presenting a good performance in other critical time industrial applications. The CAN standard is optimised for short packets and makes use of the media access control by packet priority (CSMA/AMP). The DeviceNet protocol is oriented to packets, and each packet has a specific priority that is used for moderating the access to the bus in case of a simultaneous transmission.

A CAN packet frame starts with an initial bit for synchronisation, and the next identifier performs the refereeing of the packet, in which a logic 0 dominates over a logic 1. A node that attempts to transmit a packet waits until the bus is free. Then, it starts sending the identifier of its packet bit by bit. Conflicts due to bus access are solved during transmission by a bit-level referee process, which is the initial part of each frame. With this, if two nodes attempt to transmit their packets at the same time, both keep on sending packets and listening to the network. If one of them receives a different bit of their sent bits, this loses the right to continue transmitting its packet. With this procedure, a transmission in transit has never a collision.

In a network based on the CAN standard, data is transmitted and received using a frame of packets that delivers data from a sending node to one or more receiving nodes. Data do not necessarily contain the addresses of the sender or receiver of the packet. Instead, each packet is labelled by a unique identifier within the network. All the rest of the nodes in the network receive the packet and accept or discard it depending on the configuration of filters applied to the identifier. This operation form is known as multicast.

The frame format is shown in Fig. 3.1. The complete header is 47 bits long, including the fields for the start of frame (SOF), 11 refereeing bits (identifier), control, data, cyclic redundancy check (CRC), acknowledgement (ACK), end of frame (EOF), and interruption (INT). The size of the data field is between 0 and 8 bytes.

The greatest liability of DeviceNet, compared with other network protocols, is its low transmission rate (1 Mbps maximum), which limits its performance regarding other control networks. DeviceNet also presents a low performance when the number of data items to transmit is large, although it supports data fragmentation larger than

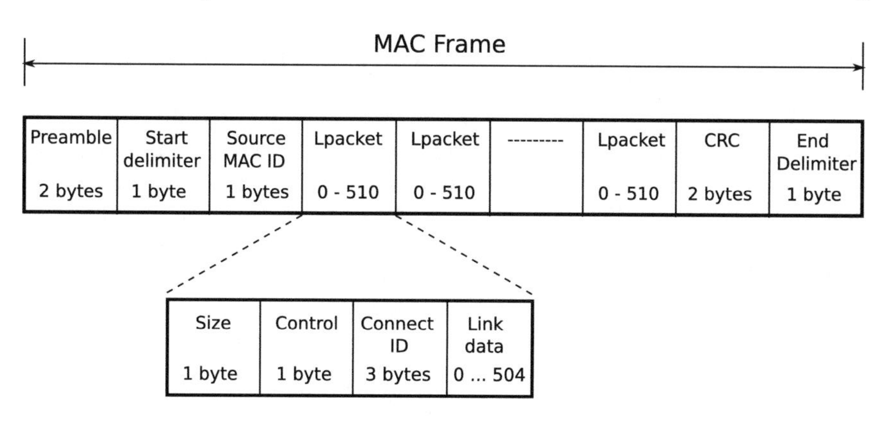

Fig. 3.2 Frame of packets ControlNet standard

8 bytes. The problem here is that, for each 8 bytes data, it is necessary to send the 47 bytes header.

In the CAN standard, data are sent in frames of packets as shown in Fig. 3.2, where a packet can be sent periodically, sporadically, or by demand. A priority is assigned to each packet, which determines its access to the network.

ControlNet. Protocols such as MAP (Manufacturing Automation Protocol), PROFIBUS (PROcess Field BUS), and ControlNet are typical examples of Token bus control networks. All these protocols are deterministic, due to the maximum waiting time before sending a packet can be characterised by the rotation time of the token. The Token bus protocol (IEEE 802.4) allows a linear, multipoint, tree, or segmented topology [11].

Regarding the topology, nodes with a Token bus protocol are arranged into a ring, and in the specific case of ControlNet, each node has the address of its predecessor and its successor. During operation, the node with the token transmits its frames until finishing them or reaching a deadline time. The node then passes the token to its successor in the network. If a node has no frames to send, it only passes the token to its successor. The physical location of the successor is not important since the token is sent to the logical neighbour. Frame collision does not occur since just one node is able to transmit at a given time. This protocol also guarantees a maximum access time for each node and generates a new token if the node that has it stops sending and/or does not pass the token to its successor. Nodes can also be dynamically added to the bus, or be retired from the logical ring.

The frame format of ControlNet is shown in Fig. 3.2. The total frame has 7 bytes, including the preamble, start delimiter, MAC source identifier, cyclic redundant check (CRC), and end delimiter, besides the data frame, also known as link packet (Lpacket). A frame may include several Lpackets, containing fields such as size, control, label, and data size with the total frame size between 0 and 510 bytes. The individual destiny address is specified in the label field. The size field specifies the number of pairs of bits (3–255) contained in an individual Lpacket, including the size, control, label, and link data fields.

The ControlNet protocol (Fig. 3.2) is a mechanism of implicit token passing and assigns a unique ID MAC (1–99) to each node. In general, the node with the token can send its data. However, there is no real token pass around the network. Instead, each node checks the source ID MAC of each received frame. When a frame finishes, each node establishes an implicit registry of the token from the source plus 1 (source ID MAC + 1). If the implicit registry of the token is the same then its own ID MAC, this node is now able to send packets. All nodes have the same values in their registries, preventing collisions in the media. If a node has no data to send, it does send it to an empty Lpacket field, known as a null trace.

The cycle length, called network update time (NUT) in ControlNet, is divided into three parts: programmed, non-programmed, or security band. During the programmed part, each node can transmit data during the programmed/critical time, in order to obtain the implicit token from node 0 to S. During the non-programmed part, nodes from 0 to U share the opportunity of transmitting non-critical data in a round-robin way until the non-programmed assignation expires. When the time for security band is ready, all nodes stop transmitting, and only the node with the lowest ID MAC, known as moderator, is able to transmit a maintenance packet, known as moderator packet, carrying out the synchronization of all timers in each node, and publishing the critical connexion parameters such as the NUT, node time, S, U, etc. If the packet moderator is not received during two consecutive NUTs, the node with the lower ID MAC starts transmitting the moderator packet on the part of the security band within the third NUT. Besides, if the moderator notices that another node has a lower ID MAC than its own, immediately cancels its role as moderator.

Token Bus is a deterministic protocol that provides performance and efficiency in high-load networks [11, 12]. During operation, Token Bus is able to dynamically add or remove nodes. This differs from Token Ring, in which nodes are a physical part of the ring, and cannot be added or removed. The programmed and non-programmed parts provide that, in each NUT cycle, the networks such as ControlNet that make use of Token Bus be able to access critical-time and non-critical-time packets.

A liability of ControlNet (Token Bus) is that, even though is efficient and deterministic at high-load traffic, if its channels have low traffic, its performance is similar to the contention protocols performance [13]. In general, when there are many nodes in a logical ring, a great percentage of network time is used to pass the token between nodes, even when data traffic is light [12]. Another disadvantage is the restriction to manage 48 nodes within the network, or 99 if there are three repeaters.

Ethernet. Another network widely used as a control network is Ethernet, which makes use of the CSMA/CD protocol (IEEE 802.3) to solve the access to the communication media. When a node attempts to send, it listens to the network. If the network is busy, the node waits until it is free. Otherwise, the node immediately sends. While the node is sending, it also listens to the network, to detect a collision. This is, if two or more listen that the network is free, and attempt to simultaneously send, the packets collide and get damaged. So, when detecting a collision, nodes

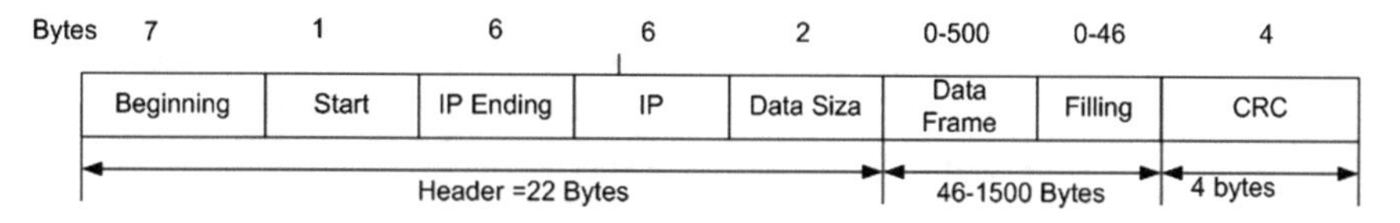

Fig. 3.3 Frame format for Ethernet

stop sending and wait a random time to attempt sending again. This random time is provided by a BEB (Binary Exponential Backtracking) algorithm, in which the retransmission time is randomly chosen between 0 and $2^i - 1$ time slots, in which i is the ith collision event detected by the node, and a time slot is the minimal time required for a roundtrip transmission. However, after detecting 10 consecutive collisions, the interval is set to a maximum of 1023 slots. After 16 collisions, the node stops transmitting and reports a processor failure. The node recovery is performed in higher communication layers [14].

The Ethernet frame is shown if Fig. 3.3. The header is composed of 22 bytes, and the end of frame is 4 bytes. The data size is between 46 and 1500 bytes. The minimum frame size is 72 bytes, including 46 bytes of data, 12 bytes for the destiny and source addresses, 4 bytes for checksum, and 8 bytes for the preamble and the beginning of the delimiter. If the size of data to be sent in a frame is less than 46 bytes, the refill field is used to complete the minimum frame data size. There are two reasons for a minimum size: first, this is easily the distinction between valid frames and garbage, since when a collision is detected, the current frame is truncated, meaning that some bits and parts of the frame are lost; second, it prevents any node to finish a short frame before the first bit reaches the end of the cable, where it may collide with another frame. Thus, the minimum packet size defines the maximum length of the cable of an Ethernet network: with a maximum length of 2500 m, and four repeaters, the time slot is 51.2 μs, which is the time required to transmit 64 bytes at 10 Mbps [14].

Due to the low overload of media access, Ethernet uses a simple algorithm for the operation of the network, and it almost has no time delays at low network loads [15]. A minimum bandwidth is used to get the access to the network, in comparison with protocols such as Token Bus or Token Ring. The type of Ethernet commonly used as a control network is Modbus/TCP, which is a 10 Mbps standard Ethernet; high-speed Ethernet of 100 Mbps or 1 Gbps is mainly used in data networks [14, 16, 17], although it is also used in control networks [14].

The most important disadvantages of Ethernet are that it is not a deterministic protocol, and it does not support packet prioritisation. For a high-load in a network, its great problem is the rising number of collisions, since this affects the performance, producing time delays that may not be bounded, and further, packet loss [11]. Another liability is the effect of the BEB algorithm, in which a node exclusively sends packets for a long time, while other nodes wait for access to the media, causing performance degradation [10]. Based on the BEB algorithm, a packet can be discarded after a series of collisions, and also, the communication from side to side is not guaranteed. Due to the requirement of a minimum valid frame size, Ethernet makes use of a large packet size even for transmitting a small amount of data.

There have been several proposals to improve the performance of Ethernet in control applications. For example, time stamping each packet before it is sent. This requires clock synchronisation, which is not easy to achieve especially in this kind of network [16]. Several schemes based on deterministic retransmission time delays for collided packages of a CSMA/CD result in a bounded time for all packets. Nevertheless, this is achieved by minimising the performance of CSMA/CD, moderating the channel utilisation in terms of time delay performance [18]. Other solutions, such as LonWorks, provide a level of prioritisation in CSMA/CD to improve time response of critical packets [19]. Using switched Ethernet, subdividing the network architecture is another way to increase its efficiency [11].

Real-time communication. In NCSs, the network becomes a resource shared by several communication entities. Thus, the management of communication resources is of special interest. Messages are exchanged between nodes. Commonly, these messages are fragmented in segments for its transmission through the network. Each segment is managed by the network as a basic transmission unit, called packet, and packet transmission is non-expulsive. Packet transmission through the network is analogous to the execution of tasks on processors; just as tasks, the kind message can be periodic, aperiodic, or sporadic. A message also has a transmission time corresponding to the worst case, and a relative deadline. Transmitting a message can be seen as a task, with the transmission time as the execution time.

Typically, networks are designed based on the 7 layers Open System Interconnection (OSI) protocol [20]. It is protocol responsibility to guarantee real-time communication. Problems such as resource management are tackled at different levels of protocol layers, combining several strategies. From the point of view of scheduling, the network is a shared resource, for which several messages race against each other. The important aim of coordinating the access to active nodes relays on the Medium Access Control (MAC) protocols [21].

Depending on how the nodes access the communication media, two MAC protocol categories can be identified [22]:

1. Controlled Access. The time and order of each node that accesses the communication media are controlled by specific algorithms, and therefore, there are no collisions. These are called contention-free MAC protocols. An example of these protocols is the Time Division Multiplex Access (TDMA), which is comparable with off-line scheduling of processors. Several Fieldbus, such as Profibus, P-NET, IEC 1158, WorldFIP, Interbus, FF, etc. belong to this category.
2. Random Access. Each node may request access to the communication media in an arbitrary, random moment. This conduces to collisions among different nodes. These MAC protocols are based on contentions. The Carrier Sense Multiple Access (CSMA) and variants are typical examples of this kind of protocols.

In the following paragraphs, MAC protocol are described in general terms, based on their description by [13, 21].

TDMA. In TDMA there is a base station that coordinates the nodes connected to the network. The network time is divided into segments of a fixed size. Each node has assigned a certain number of segments on which it can send or receive messages. Thus, multiple nodes are allowed to access a shared network without collisions. TDMA is deterministic, which is suitable for real-time applications. However, due to its inflexibility, TDMA is not used in data applications in which the traffic is undefined, such as the Internet.

CSMA/CD. This protocol is based on CSMA, but with collision detection. It consists of a set of rules to determine how nodes respond when there are collisions. The term "Carrier Sense" refers to the fact that several nodes listen to the network before trying to send. If the network is busy, they wait until the network becomes available. "Multiple Access" means that more than one node can sense (listening and waiting to send) at every moment. "Collision Detection" refers to when multiple nodes accidentally attempt to send at the same time, and a collision can be detected.

CSMA/CD is an IEEE 802.3 and ISO 8802.3 standard. Ethernet networks make use of CSMA/CD to solve contention conflicts in the communication media. In Ethernet, any node can try to send a packet at any moment. If the network is available, then the sending node starts to transmit. While it is transmitting, it listens to detect whether there is a collision. If a collision occurs, each node that participates in it stops transmitting and waits for a random time interval for re-transmitting. This random time is determined using the BEB algorithm.

The main advantage of CSMA/CD is that it is easy to implement, which has helped to Ethernet to become the most important standard in local area networks. However, CSMA/CD is a protocol non-deterministic, and it does not support any message prioritisation. The collisions are the greatest problem in high-load networks, due to the resulting time delay cannot be bounded.

CSMA/CA. This protocol means CSMA with "Collision Avoidance". This protocol is another modification of CSMA, in which a collision is avoided, instead of detected. This protocol is normally used to improve the performance of CSMA. The design reduces the probability of collisions when using multiple nodes sharing a communication media, where collisions occur. This is actually the collision avoidance. Differently, from CSMA/CD, which provides a treatment only after detecting a collision, CSMS/CA acts to avoid collisions before they happen. CSMA/CA is commonly used in applications where CSMA/CD cannot be implemented, due to the nature of the communication media. A typical example is the protocol IEEE 802.11, which specifies two functions that coordinate the access to the media: the Distributed Coordination Function (DCF), based on CSMA/CA, and the Point Coordination Function (PCF). In contrast with wired nodes, wireless nodes cannot detect collisions, since they are half-duplex, this is, they cannot send and receive signals at the same time. This is the reason for which CSMA/CD is not applicable to wireless communications. In IEEE 802.11, each node senses the media before starting a transmission. The Carrier Detection is performed by one of two mechanisms: one physical, and another virtual. The time lapses are divided into multiple frames or segments, and the time interval between segments is known as Inter-Frame Space (IFS). Different IFSs are defined to provide priority levels for the access in wireless media.

If the media is busy at the sensing moment, the node waits for the end of the current transmission, and starts its contention, called a "backoff process". The node selects a random backoff time given by $rs = random()$, which provides a pseudo-random integer from a uniform distribution over the interval $[0, VC]$, where VC is the contention window, and it is an integer within the range of values of the features of the physical layer VC_{min} and VC_{max}, this is $VC_{min} < VC < VC_{max}$, and the time lapse is that correspondent to the physical layer. VC exponentially increases as the number of retransmission attempts of a packet increments. During the backoff process, the backoff timer decreases in terms of the segment as long as the media is not busy. When the media is busy, this timer stops. When backoff time concludes, if the network is still not busy, the packet is sent. The node which has the shortest contention time delay wins, and sends its packet. The other nodes just wait for the next contention at the end of sending the current packet. If another collision occurs, a new backoff time is selected, and the contention procedure starts over until at a certain moment the limit is exceeded. Since contention is based on a random function, and it is used for each packet, each node is given the same opportunity to access the communication media. Once an integral packet is received, the receiving node waits for a IFS interval, and transmits a received acknowledgement (ACK) back to the emitter node, which indicated a successful transmission. If the emitter node does not receive an ACK, it assumes that the packet has collided, and attempts to retransmit, entering again to the backoff process. To reduce the possibility of collisions, the VC is doubled until it reaches the higher limit VC_{max}, after each retransmission attempt. VC is then restored to the value of VC_{min} after each successful transmission. But when an error occurs, a packet has to be retransmitted by the emitter node.

CSMA/CA delivers a better service, offering guarantees in bandwidth and time delay terms. Just as CSMA/CD, this is a non-deterministic protocol, but it has several advantages. For example, CSMA/CA is suited for network protocols such as TCP/IP, adapting quite well to variable traffic conditions, and it is very robust against interference.

CSMA/BA. This protocol is called CSMA with Bitwise Arbitration. It is a protocol commonly used by Controller Area Network 3 (CAN 3) and DeviceNet to solve collisions. This protocol is sometimes considered as a special case of CSMA/CD or CSMA/CA, and also referred as CSMA/AMP (CSMA with Arbitration of Message Priority). Just like CSMA/CD and CSMA/CA, each node in the network using CSMA/BA has to wait until an inactive period finishes before starting to send a message. Once this period expires, each node is able to send its message. Bitwise Arbitration means that collisions are solved by an identification bit, based on the pre-programmed priority of each message in the identifier field. For CAN, the identifier is a single sequence of 11 or 29 bits, which assigns a priority to the message, and allows the receiver to filter messages. Due to CAN is a broadcast network, messages do not contain explicit addresses of the receiver or the sender.

The arbitration mechanism CAN works as follows: a binary number of ID message with a small value indicates a higher priority. A CAN message associated with the highest priority wins arbitration. This is achieved by transmitting messages using a binary model of dominant and receive bits, where a dominant is a logical 0, and a receive is a logical 1. In CAN bus, a dominant bit always overwrites a receive bit. If multiple nodes simultaneously transmit, and a node sends a 0, then all nodes monitoring the network read a 0. If all nodes transmit a 1, all nodes read such a 1. CAN behaves like an AND gate. During the transmission of the identifier field, if a node sends a 1 and reads a 0, this means that there has been a collision with at least one message of a higher priority, and consequently, this node aborts transmission of its message. The message with the highest priority is sent, and it proceeds without perceiving any collision until successfully finishing transmission.

The arbitration procedure CAN is non-destructive, in the sense that the winner node just continues transmission of its message without being destroyed or corrupted by any other node. The determinism of CAN makes it particularly attractive for use in Real-time Control Systems.

Due to its conception, CAN protocol has earned popularity in the automation industry, medical control, automotive applications, etc. The most important disadvantage of CAN compared with other networks is its slow transmission rate. This is mainly due to the Bitwise Arbitration mechanism, which imposes strict limitations over the physical features of the network, including length and data transmission rate. Because of this, transmission rates larger than a 1Mbps are only possible if the length of the network is shorter than 40 m. For networks with a larger length, the data transmission rate decreases to 125 kbps for 500 m.

3.3 Distributed Systems Changing Methodologies

Nowadays, distributed systems tend to be inherently heterogeneous in terms of communication devices as well as processors. Therefore, in this section, a strategy capable of dealing with different communication media to model the effects of this appearance, is introduced. In such a way that the scheduling strategy is based on modifications of frequency transmission of individual nodes in the system [23]. Data exchange is carried out under specific frequency transmission, through a scheduler node, which applies a control frequency algorithm, based on the information retrieved by smart sensors. This particular type of sensors is distinguished by its communication and consensus capabilities [24]. Hence, the main objective here is to expose the importance of managing sampling rates, and an algorithm to modify them to obtain an improvement in the scheduling strategy. This is demonstrated using a simulated case study, based on two degrees of freedom (2-DOF) helicopter simulation benchmark [25], in Chap. 6. This simulation provides an approximation to system response, in which, for demonstration purposes, the main results are obtained for a typical fault scenario [26]. Thus, for this simulation, two scheduling strategies are implemented using TrueTime [26, 27]: one showing a dynamic scheduling, and the other carrying out static scheduling. Both strategies execute on a real-time distributed system, modifying the frequency of transmission, as well as the periods of tasks in individual nodes. Thus, due to network use, both strategies impact on the quality of

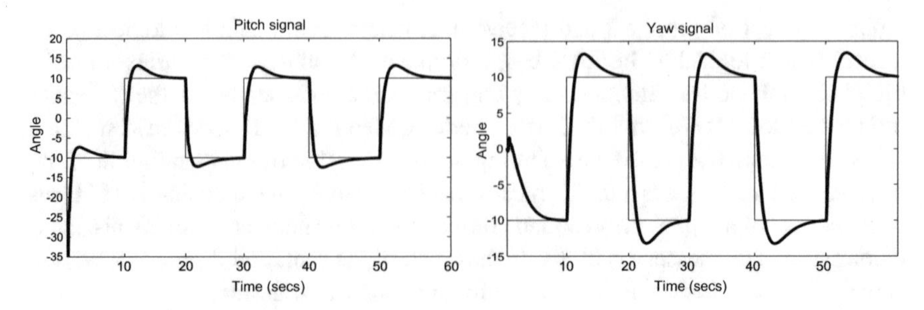

Fig. 3.4 Stable behaviour of the RTDS

performance of the NCS [28]. The second strategy modifies the periods of tasks, and network access is assigned through a static scheduling algorithm; the first strategy achieves dynamic schedulability by controlling the rate of frequency transmission into regions of stable frequencies. This region is bounded by a minimum and a maximum transmission rate. The objective is to expose the advantages of using dynamic scheduling in an ad-hoc implementation.

The first scheduling strategy consists of determining a base period, obtaining the operational periods of all nodes that participate in the RTDS. Even though all the nodes can use this base period as their operational period, the strategy includes the option of having different periods for each node, by means of the base period and a dispersion factor, namely hyperperiod which is discussed later in this book. The base period is the operational period for the controller periodic task in the corresponding controller node. The base period and dispersion factor are used to obtain the actual sampling periods of a number of sensor nodes.

Reference [29] provide some tests to quantify the performance of the NCS based on the IAE as a metric. The base period and dispersion parameters are fixed in specific values. These values are used to evaluate such a performance when all nodes are competing to get the network access. According to the network access control algorithm, the system tends to be unstable, since the IAE increases.

Figure 3.4 shows a typical, stable behaviour of the RTDS for the 2-DOF Helicopter simulation, using a base period within the interval [0.015, 0.020] seconds, and dispersion fixed to 0.15 for all nodes. The simulation is carried out during 50 s.

Notice that, however, when the base period is out of the proposed interval and/or the scheduler does not assign a proper bandwidth, the system easily becomes unstable. Figure 3.5 shows such unacceptable behaviour. The access to the network is defined using a scheduling algorithm, executing on the scheduler node as a periodic task. The scheduler generates an *id*, which each sensor uses to access the network, in order to send data to the controller.

Figure 3.6 shows the communication between sensor and controller nodes, supervised by the scheduler node. For this, each sensor node is aware of its minimum and maximum frequency rates (f_m, f_x), and based upon messages sent to the controller, it could estimate its own real transmission frequency rate (f_r). Thus, let a RTDS with

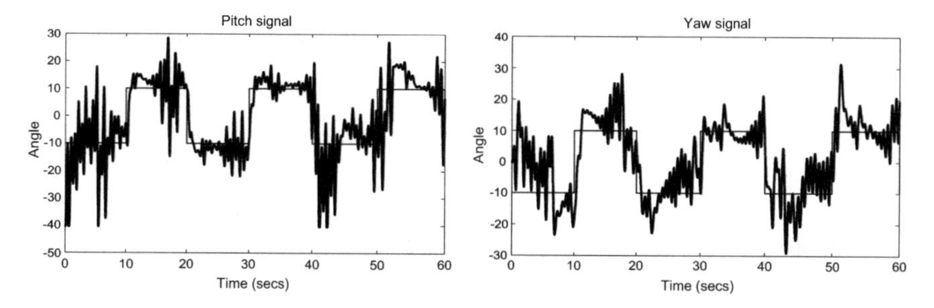

Fig. 3.5 Unacceptable behaviour of the RTDS

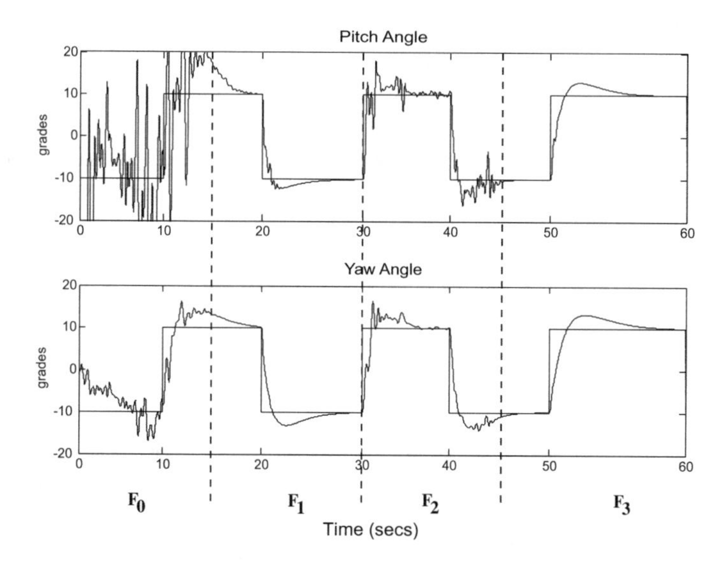

Fig. 3.6 Communication between sensor and controller nodes

k nodes be such that each node performs n tasks, and each task t_i has a period p_i and a consumption c_i.

Since an optimum fixed-priority scheduler has an upper bound of processor utilisation [30], it is necessary to consider each processor that uses the network in the RTDS:

$$U = \sum_{i=1}^{n} \varepsilon_i * f_x^i \leq 1 \tag{3.4}$$

where ε_i is the task consumption time and f_x^i is the maximum frequency.

For the actual purposes, let us assume that there are k nodes, each one having n tasks, and there is a set of possible task configurations that satisfy Eq. 3.4. For example, consider the set H_j^*, which contains all task configurations corresponding to those parameters in which all n tasks of each processor j are scheduled for all s configurations:

$$H_j^* = \left\{H_j^1, H_j^2, \ldots, H_j^s\right\} \tag{3.5}$$

where each element of the set $\{H_j^1, H_j^2, \ldots, H_j^s\}$, is a local schedulable subset. Thus, a global schedulable set is which comprises all k processors of the RTDS, and hence, the overall RTDS is schedulable.

$$H = \left\{H_j^* \middle| j = 1, 2, \ldots, k\right\} = \{H_1^*, H_2^*, \ldots, H_k^*\} \tag{3.6}$$

The RTDS can also be modelled as a linear time-invariant system, in which the state variables $\{x_1, x_2, \ldots, x_n\}$ take the values of the desired transmission frequencies. Let us assume that there are several ratios between real transmission frequencies $\{f_1, f_2, \ldots, f_n\}$, and the external input frequencies $\{u_1, u_2, \ldots, u_n\}$ [23]. These ratios serve as coefficients of the matrices A and B of the discrete time linear system:

$$x[k+1] = Ax[k] + Bu[k] \tag{3.7}$$

$$y[k] = Cx[k] \tag{3.8}$$

where $A \in \Re^{n\times n}$ is the matrix of ratios between frequencies of all nodes, $B \in \Re^{n\times n}$ is the scale frequencies matrix, $C \in \Re^{n\times n}$ is the matrix with ordered frequencies, $x \in \Re^n$ is the desired frequencies vector, and $y \in \Re^n$ is the output frequencies vector. The input $u \in \Re^n$ is a vector of reference frequencies of the nodes in the RTDS. Let $a_{ij} \in A$ given by a function φ of minimal frequencies f_m of node i, and $b_{ij} \in B$ given by a γ function of maximal frequencies f_x, so that:

$$a_{ij} = \varphi\left(f_m^1, f_m^2, \ldots, f_m^n\right) \tag{3.9}$$

$$b_{ij} = \gamma\left(f_x^1, f_x^2, \ldots, f_x^n\right) \tag{3.10}$$

The purpose of the controller is to regulate the transmission frequencies of each of the nodes present in the system to guarantee both the transmission frequency of each node and minimise the time delays of the system to disturbances. The control input is given by the error between the current frequencies f_r and f_d:

$$f_d = \left[f_d^1, f_d^2, \ldots, f_d^n\right] \tag{3.11}$$

$$f_m = \left[f_m^1, f_m^2, \ldots, f_m^n\right] \tag{3.12}$$

$$f_r = \left[f_r^1, f_r^2, \ldots, f_r^n\right] \tag{3.13}$$

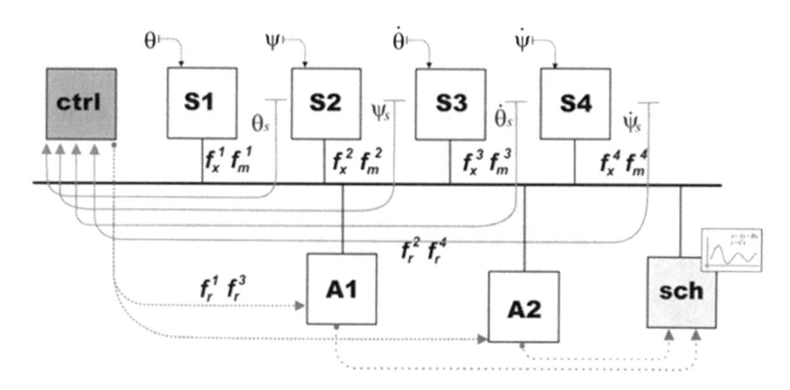

Fig. 3.7 Data transmission between sensor and scheduler nodes

The control input is given by a function of desired frequencies (f_d) and real frequencies (f_r) for all the sensors:

$$u = -k^s(f_d - f_r) = -k^s e(k) \tag{3.14}$$

where k^s is the control gain defined using the LQR algorithm. Matrices A, B, and C are dimensionally correct with the schedulability restriction expressed in Eq. 3.4. Defining $x(k) = (f_d - f_r)(k)$, and substituting Eq. 3.14 in Eq. 3.7:

$$x(k+1) = A(f_d(k) - f_r(k)) + B(k^s(f_d(k) - f_r(k)) \tag{3.15}$$

$$x(k+1) = (A - Bk^s)(f_d - f_r)(k) \tag{3.16}$$

Figure 3.7 shows the data transmission from a sensor node to a scheduler node, sending frequency data through the network. In this case, four sensors, two actuators, and one controller compose the distributed system, based on the simulation of the 2-DOF Helicopter, as presented later in this chapter. Each time, the scheduler node uses frequency information to obtain a new frequency transmission, aiming for an efficient utilisation of the network. This description is widely discussed in Chaps. 4 and 6.

3.4 Dynamic Representation of Distributed Systems

In order to study the impact of network utilisation on closed control loop, the 2DOF Helicopter control model is built as a RTDS [31], following the strategy presented in Sect. 3.3. Several nodes are connected through a common communication network. The experiments focus on network scheduling, and the main objective is to balance the amount of data sent through the network, in order to avoid latency and undersampling.

Two network scheduling proposals are given to explore several aspects of control performance when the use of the network exceeds network bandwidth.

To have a performance index, and thus, quantify the system's performance, the integral of the absolute value of the error IAE is used:

$$IAE = \int_{t_0}^{t_f} |e(t)|\, dt \approx \sum_{k=k_0}^{k_f} |r(kh) - y(kh)| \tag{3.17}$$

where $r(kh)$ is the reference signal, $y(kh)$ is the system output signal, $t_0(k_0)$ and $t_f(k_f)$ are the minimum and maximum times of the evaluation period.

A fundamental requirement of a control system is the stability. The control community has several kinds of stability. For example, a system is stable if, for any bounded input, the output is bounded; an unstable system gives an unbound output. Apart from stability, the transient behaviour is another focus of attention for control systems design. The control error $e(t)$, which is defined as the difference between the setpoint $r(t)$ and the system output $y(t)$ is computed using the IAE. This value gives an index to evaluate the performance of the system. For this index, a large value corresponds to worse performance. Here, stable behaviour deals with small values of the IAE and unstable behaviour deals with large values of the IAE.

The particular RTDS used in this section for the experiments consists of 8 processors. These real-time kernel processors and the network are simulated using TrueTime [26]. The network used is a CSMA/AMP (CAN) with a transmission rate of 80000 bits/s, a minimum frame size of 40 bits, and no data loss. The TrueTime Network block simulates the physical and medium-access layer of various local-area networks. Other types of networks supported by TrueTime are CSMA/CD (Ethernet), CSMA/AMP (CAN), Round Robin (Token Bus), FDMA, TDMA (TTP), Switched Ethernet, WLAN (802.11b), and ZigBee (802.15.4). Notice that the network blocks only simulate the medium access (the scheduling), possible collisions, or interference, as well as point-to-point/broadcast transmissions. Higher-layer protocols such as TCP/IP are not supported but may be implemented as processes in the block nodes.

Four sensor nodes execute periodic tasks to sense control signals, as well as other additional periodic tasks. Each task has a period pi and time consumption ci (Fig. 3.8). The sensed control signals are $\left(\theta, \psi, \dot{\theta}, \dot{\psi}\right) \rightarrow \left(\theta_d, \psi_d, \dot{\theta}_d, \dot{\psi}_d\right)$. This model has a controller node, depicted on the left side (Fig. 3.8). This controller takes the control law from the LQR module using a task, which activates by event [31]. The time consumption of the controller task is the maximum average time it takes to compute the control law. The controller node uses the values from sensors, and sends control outputs u_p and u_y, that correspond to the pitch and yaw actuator voltages. Two actuator nodes, located on the bottom right corner, receive signals from the controller node. Finally, the scheduler node, located on the top right corner, organises the activity of the other seven nodes, and it is responsible for periodic allocation bandwidth, used by these nodes.

Each node initializes specifying the number of inputs and outputs of the respective TrueTime kernel block, defining a scheduling policy of the processor, and creating

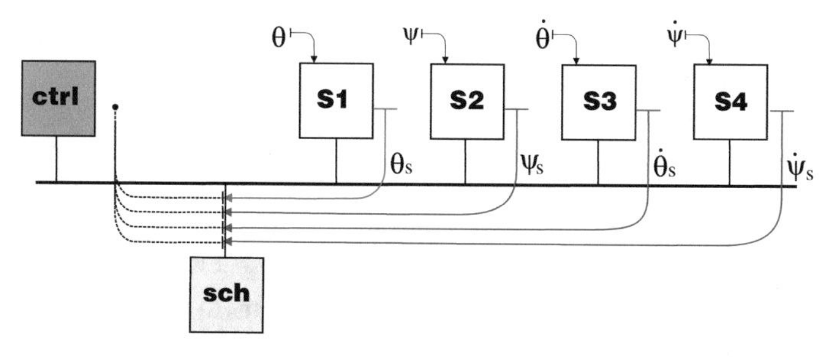

Fig. 3.8 RTDS into the Helicopter model

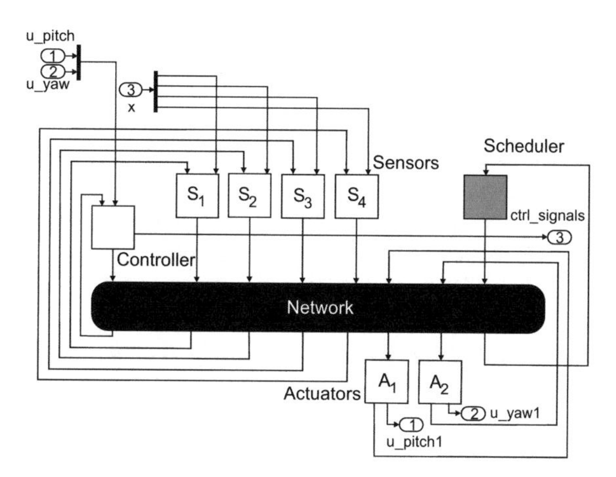

Fig. 3.9 NCS for the Helicopter model

periodic tasks for the simulation. These tasks involve parameters about the periodic times and the consumption times. The task periodic times define the time interval between tasks, whereas the consumption times refer to the execution time of the task. Figure 3.9 shows the 2-DOF Helicopter model, with a RTDS, where feedback control loop is closed through a communication network.

Changes in the real-time task parameters of the RTDS commonly impact on network utilisation, and therefore, on the control performance [13, 32]. The problem to tackle, thus, is to find a proper way to schedule the common communication network of the RTDS, based on managing a feasible sampling period, capable of keeping both, the network load and required integrated performance.

Here, the NS2 network simulator is not applied, since the use of NS2 has been reviewed in terms of network bandwidth, and some scheduling strategies such as EDF or Token Ring [5]. From this, several strategies may be usable for scheduling. However, none of them guaranteed a valid response in terms of time constraints, since real-time is an important restriction for the purposes here. Even though NS2 allows simulation of complex situations, in terms of data exchange, it neglects the local jitter that is inherent in the simulated transactions, since it makes use of a scale factor that is quite significant compared to the delay of data transfer. Hence, the accuracy of NS2 simulator is out of the scope of this approach, since the work here necessary deals with modelling the features of the data exchange.

Since its creation in the mid-70s, Distributed Artificial Intelligence (DAI) has quickly evolved and diversified. Nowadays, it is a research and application field that gathers results, concepts, and ideas from several other disciplines, including Artificial Intelligence (AI), informatics, sociology, economics, science and philosophy, etc. Its multidisciplinary nature makes it difficult to precisely and briefly characterise DAI. The following definition has been used as a starting point to explore this field, and it is taken here as an initial reference [33]:

> DAI is the study, implementation, and application of multiagent systems, this is, systems in which several intelligent entities interact, pursuing a set of objectives or perform a set of tasks.

An agent is a physical or computational entity that perceives and acts over its environment [33]. It is autonomous in the sense that its behaviour depends on its own experience (at least, partially). It is an intelligent entity that operates in a flexible and rational way in a variety of environmental circumstances, given its perceptual capacity. The behaviour flexibility and the rationality are achieved by the agent based on primary processes, such as problem solver, scheduling, decision making, and learning. As an interactive entity, an agent can be affected by its activities or by other agents' activities, and perhaps by human beings. A key interaction pattern in multiagent systems is the objective-oriented coordination, along with the task-oriented coordination [34]. Both are effective in cooperative and competitive situations. In the case of cooperation, several agents attempt to combine efforts to achieve as a group what each isolated individual cannot. In the case of competence, several agents attempt to get what only one of them can have. The objective in the long term of DAI is developing mechanisms and methods that allow agents to interact with humans, and understand the interactions among intelligent entities, whether they are informatics, human, or both. This objective proposes some challenges, all of them around the elemental question of "when" and "how" to interact with "who".

For Bond [35], and Huhns [36], there are two basic reasons that justify the use of DAI, and have served as main impulses behind the growth of this field since two decades:

1. Multiagent systems are able to play the main role in computer science and its applications, not only nowadays, but also in the future. The current computing platforms and information environments are distributed, open, and heterogeneous. Computational systems are not isolated; they are connected among themselves and their users. The growing complexity of the information systems and of the information itself is observed as well in their applications. This frequently goes beyond the level of conventional, centralised computing, since it requires, for instance, processing large amounts of data. Computers have to act more than individuals than as only parts; this holistic focus is being applied nowadays for modelling dynamic systems, and by the kind of real-time scheduling [37]. The DAI provides a technology that manages the high-level interaction and the complex applications for computing in a modern information system.

2. Multiagent systems are able to perform an important role in the development and analysis of models and theories of interactivity in human societies [38, 39]. Human beings interact in diverse ways and at several levels: for example, observing and modelling each other, requesting and offering information, negotiating and discussing, developing a shared vision of their environment, detecting and solving conflicts, creating and dissolving organisation structures such as teams, committees, and economies. Many interaction processes among human beings are yet not well understood, even though they are an integral part of our daily life. The technology of DAI possibles to explore the sociological and psychological foundations of human relations.

In [40], it is required that productive systems have to be more flexible for carrying out different actions that improve their capacity, and with such an improvement, achieve their objectives. From this, their complexity has increased. Traditionally, this kind of systems makes use of centralised control, which has been modified towards distributed control. Distributed systems, as stated before, consists of the division of complete systems into self-managed parts or modules that interact among themselves. These interactions rule the function of the elements, on which the control of the system is based. Traditional control systems are hierarchical, are applied in cases of little flexibility and simple production sequences. However, a few years ago, a turn has been taken to the distributed control system based on dealt methods. These new systems are divided into different components that constitute autonomous execution units of strategies, provided with their own control capacities. The main properties of these components are autonomy and cooperation.

Agents. A new paradigm has been proposed based on agents that goes beyond the object-oriented methodologies. Agents constitute improved objects in the sense that objects are passive and do not activate a behaviour. Instead, agents are capable of establishing relations with other agents. These relations are not foreseen during the build-time of agents, and they are capable of adapting all correspondent rational features to the environment [40]. The relations among agents do not follow a hierarchical structure. They are heterarchical, this is, they include processes of negotiation in their interactions with other agents, since there are no restrictions on levels, agents negotiate among themselves and depending on the needs of the system [41]. Agents are more adaptable and flexible than hierarchical systems. The heterarchical structures represent the most relevant challenge in the design of multiagent systems. In such structures, hierarchical organisations are eliminated, or at least, restricted. Cooperation and competence among agents should be replaced by supervision and rank orders [41].

There are several classifications of agents which depend on the behaviour of each one of the agents as part of the system [41]:

- Reactive agents. These agents receive information of the environment through sensors, and after analysing this information, they consequently act to achieve their objectives. When executing their actions, the agent only needs to take several readings on the sensor, and compare them with the conditions of that instant, and

hence, execute the adequate actions. If more than one state condition of the system matches with the readings, this gives way to a reaction. Each condition is associated with a list of conditions with less weight, which it inhibits. Therefore, an action is only executed if there are no other conditions which inhibit the condition that produces it.
- Deliberative agents. These agents are considered as proactive. They work by means of logical deduction and assume the environment state through a database of formulas and logical predicates. Decision making is modelled through a set of deductive rules. The liability of these agents is that decision making is based on assuming static environments. To deal with this, Wooldridge and Jennings [39] propose the concept of deliberative agent BDI (for beliefs-desires-intentions), which has a charge of own beliefs, desires, and intentions.
- Hybrid agents. These agents combine reactivity and proactivity. A hybrid agent is autonomous, reactive, proactive, and besides, capable of establishing social behaviour. Their liability is that they are difficult to design, so they coordinate different categories for a coherent functionality. The problem is that it is complicated to determine up to what point the agent has a reactive character, and from which point it has a proactive character.

Multiagent systems. Currently, there is an attempt to build sets of autonomous and intelligent entities that cooperate to develop something, and whose communication is possible by means of mechanisms for sending and receiving messages. The tendency is to provide solutions to problems such as agent cooperation for an objective: generating subtasks and assigning them to a group of agents capable of performing such subtasks, as well as developing parallel languages and algorithms [42]. Today, developments based on several agents have had two venues: robotic agents and computational agents. This has conducted to strong distinctions regarding the dynamic and kinematic characteristic inherent to robotic agents and computational agents, that should be separately treated to define techniques of coordination and collaboration on each step [42].

The requirements demanded to a multiagent system (MAS) working in real environments are as follows [42]:

- Acquire behaviour according to the circumstances.
- Take adequate decisions at the occurrence of unforeseen events in the environment.
- Perform tasks with efficiency and in real-time.
- Be conscious of the existence of other agents in the environment.

Even though MAS has provided results in pursuing some control objectives, architectures based on negotiation among independent systems present the liability of depending excessively on the quality of the information and the means by which they communicate.

Multiagent interaction model. To manage the complexity of systems, it is proposed to use structures and techniques that decompose the problem into smaller, manageable problems, which yet may be susceptible to be further divided into yet smaller problems. This way, the solution of the problem is the composition of all the solutions to all smaller problems obtained from decomposition. Agent-oriented systems constitute an alternative to design distributed control systems. For this, it is required to decompose the problem into several autonomous entities, able to act and interact in a flexible way to achieve their objectives, and able to control themselves and their own actions [43]. Systems based on agents provide a decentralised solution, based on partial and local views from the dynamic environment, and provide to the system a high degree of flexibility and soundness. Agents are the identifiable entities with defined objectives, placed into a particular environment.

Delchamps [44] states that a collaborative MAS is based on the division of the system into small entities that encapsulate individual details and associated complexity of each one, so the computations required per involved function within each entity are achieved by means of collaboration among them. The entities that compose the system have a partial view that covers the scope in which it performs, based on its basic function. Nevertheless, the system reacts as desired if all actions are coordinated. In distributed systems, the complexity of the problems to solve demands local points of view, given that they are too large to be analysed as a whole. Solutions based on partial views frequently offer flexible and practical solutions. This way of locally thinking constitutes a promising alternative to solve complex, large-scale problems.

Ramirez [45] states that a way to combine a set of diverse software modules is to connect them regarding the requirements of data flow. This way of communication is the right choice when the set of entities, as well as the communications among them, are static. In the case that the interactions among modules are dynamic, and it is not determined at the moment that a module appears or disappears, then it is necessary to use private communication protocols. This last case makes use of the Blackboard (BB) architecture. BB architecture is composed of three types of components [43]:

- A knowledge source (KS) that works like a dual condition-action.
- The blackboard, which is a global repository of data, containing input data, possible solutions, partial solutions, and final solutions. Each modification of the blackboard may provide new KSs.
- The controller, which performs the selection of the KS that interacts and conducts to a solution.

The BB contains all possible alternatives to reach a solution. These alternatives are selected through the competencies of each agent and a KS [44], in which information is stored aiming to find better solutions.

Based on a MAS that makes use of a BB architecture, several types of communication are available [44], depending on the disposition of the BB. A particular case, the communication between agents is carried out by sending and receiving messages from all or some of the agents.

Different types of interaction between agents can be obtained regarding the disposition of the KS, the BB, and the controller. This provides a flexible implementation of the MAS for each particular case study.

Menendez and Benitez [29] propose a scheduling strategy for a RTDS, in which the base sampling period of the system ρ is determined, and with this, the operational periods of all nodes participating in the NCS are obtained. Although all nodes can use the base period as their operation period (Sect. 5.1, the strategy proposes the option of having different periods for each node, using the base period and a dispersion factor λ. This factor works as a shifting percentage between tasks executing on a node and the base period. The base period is then the operational period for the periodic tasks of the controller, executing in the corresponding control agent. Both, the base period ρ and the dispersion factor λ are used to obtain a current sampling period of the sensor components, with the following equation:

$$\begin{aligned} p_{s1} &= \rho\,(1+\lambda) \\ p_{s2} &= \rho\,(1-\lambda) \\ p_{s3} &= \rho\,(1+1.1\lambda) \\ p_{s4} &= \rho\,(1-1.1\lambda) \end{aligned} \tag{3.18}$$

In [29], the prototype of a Quanser helicopter is used as a case study, and obtain the value of the ranges of λ from 0 to 20%. For the value of 0%, all the sensors have the same period of the controller. When the dispersion is 20%, the sensors have operational periods with 20 and 22% above and below the value of the base period ρ. For values of dispersion above 25%, there is an acute loss of performance in the system. The results presented make use of the IAE value as a performance index of the system.

Hence, in [29] the following scheduling is proposed: the scheduling node executes a periodic task with period p_{sch}. The selecting node assigns a time frame to the chosen node with a certain probability distribution. The node identifier (*id*), chosen by the scheduler, is written in a shared memory. Each node ready for sending checks this value, and if it finds its *id* in such a shared memory, it sends; otherwise, the node waits one or more time lapses p_{sch}. Figure 3.9 shows an example with four sensors, from which the signals ($\theta_0 = 0$, $\psi_0 = 0$, $\dot{\theta}_0 = 0$, $\dot{\psi}_0 = 0$) corresponding to the pitch and yaw angular position, the pitch and yaw angular velocity, and the transmission of all these data to the controller, which is supervised by the scheduler node. The blocks marked as $S1$, $S2$, $S3$, and $S4$ represent the sensor agents that send data to the controller agent *ctrl* under the supervision of the scheduler agent *sch*, which assigns a transmission rate of data from sensors to the controller.

Using this scheduling strategy, the controller agent obtains 1/3 of the bandwidth assignments, while the rest of the nodes share the remaining 2/3. This procedure is feasible since it assumes the RTDS as a dynamic process in which time delays are bounded and invariant in time. Off-line, an analysis for adequate transmission period selection is performed, in order to determine the specific scheduling parameters that do not generate large time delays. Figure 3.10 shows the network activity for transmissions between the sensors and the controller.

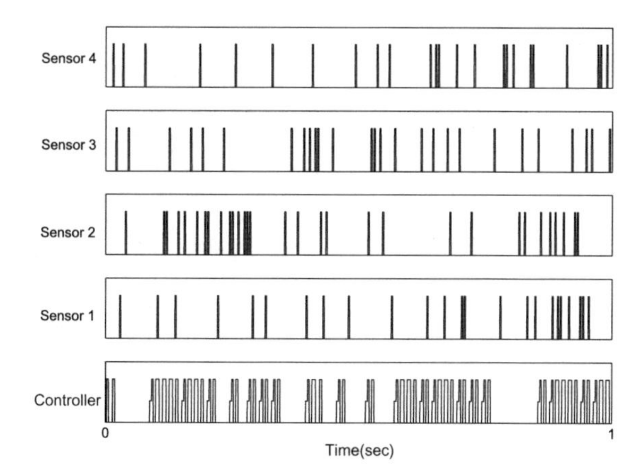

Fig. 3.10 Example with four sensors of the proposed scheduling

Reconfiguration Proposal

The management of sensor redundancy based on signal consolidation and fault masking is a technology used and proven in aircraft flight control [46]. Recently, the implementation of redundancies is carried out using intelligent devices, defined as peripheral elements with capacities for self-diagnosis, self-compensation of perturbations, and digital communications [24, 47]. These intelligent devices are basically sensors and actuators.

The scheduling strategy here proposes making use of intelligent devices as agents aiming for implementing a MAS, in which the communication between entities generates information exchange for decentralised decision making. Each agent is able to perceive its environment and cooperate with other agents. The intelligent devices are hardware entities that power up software, and that are used to perform decision processes. Currently, the use of smart elements allows the integration of control functions [24, 47]. If desired, agents may obtain and manage control information, perceive environment events, and react depending on these during system execution. The objective is that agents can communicate and collaborate among themselves to decide a reconfiguration plan, based on scheduling parameters.

In the context of the reconfiguration model, it is proposed that, instead of allowing the agents to detect and identify system faults, each sensor works as an information source for reconfiguration. Let us assume that sensors count with a mechanism for detecting and identifying faults, and thus, let us focus on the fault management by means of reconfiguration. With the fault information that each agent has, and through agreement, the kind of reconfiguration is decided. Each reconfiguration scheme used for certain faults is based on an off-line analysis. The set of all local information composes the global vision of the states of the system.

The reconfiguration system is based on a scheduling agent which modifies the scheduling parameters, based on the management of bandwidth assignation of the used network. From the reconfiguration scheme proposed by [48, 49], the steps for reconfiguration based on agents are:

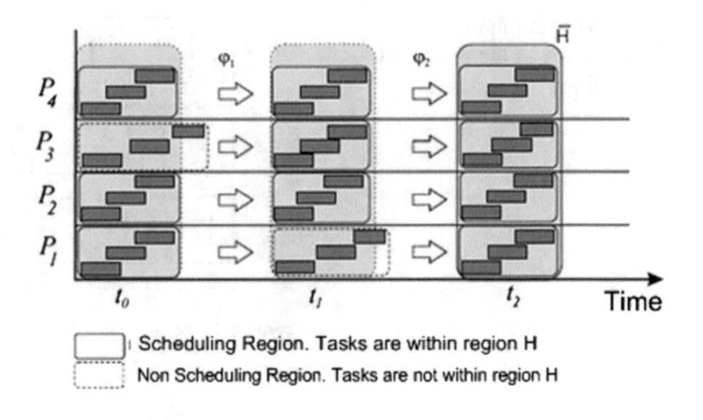

Fig. 3.11 Reconfiguration process

1. Detection and identification of the fault.
2. Share information to agree on a reconfiguration model.
3. Find an action sequence for reconfiguration that guides the system to new equilibrium states.
4. Execute the reconfiguration (time bounded) to re-stabilize the plant to a new equilibrium.

Figure 3.11 shows the reconfiguration process, in which each individual agent interacts with the others, not only in its control action but also in its reconfiguration task. The local view of each agent over what happens with the system is shared with other agents, so integrating a global knowledge of the system's functionality.

Let us consider a NCS with sensor redundancy. Each node is an agent that performs some of the tasks in closed loop: control, sensing, or actuation. For m state signals of the system, there are $B_1, B_2, B_3, \ldots, B_m$ blocks of sensing with r_i sensing agents each one. Let us denote $a(i, j)$ as the j agent of block i that performs readings of the signal x_{ij} for $i \in (1, \ldots, m), j \in (1, \ldots, r_i)$.

The fault tolerance mechanism consists of using triple sensor redundancy [50, 51], aiming to mask the fault through a voting algorithm that referees faults produced in the sensors. Based on the MAS previously described, the redundant sensors allow confederate groups of sensors that read the system states and work as agents for identifying faults (a_{fi}). The scheduling agent works as a reconfiguration agent (a_{fr}) that executes the change of the scheduling parameters, based on a previous agreement. Each agent a_{fi} is able to perform the following activities (Fig. 3.12):

- Read signals of the system state.
- Identify system faults.
- Communicate with other agents.
- Agree on the kind of reconfiguration depending on the present fault.

Each sensor agent $a_{i,j}$ executes a periodic task with period p_i and execution or consumption time c_i, and performs the activities described above. For each np ($n = 0, 1, 2, 3, \ldots$) sampling periods, there are a value of the measured signal $x_{ij}(n)$ and a value $f_{ij}(n)$ that classifies the possible fault. Notice that detection and identification

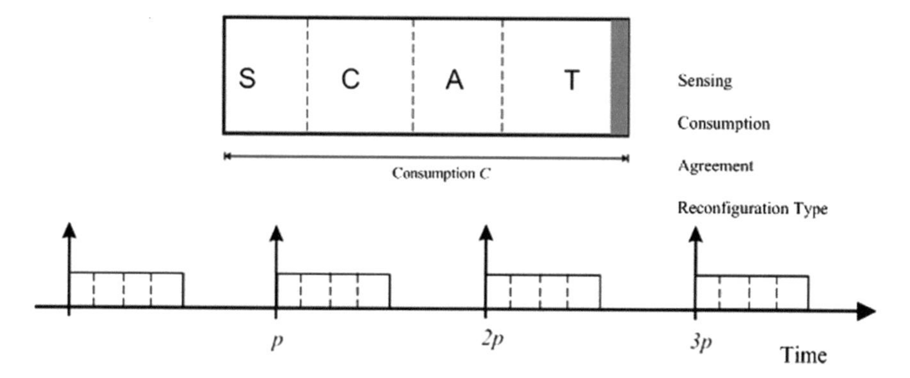

Fig. 3.12 Reconfiguration process

of faults are beyond the scope of the present book. However, it is assumed that the system is provided with a mechanism for detecting, identifying, and classifying faults present in data reading. With the previous knowledge of the effect of faults on sensors, it is possible to design a reconfiguration plan for each one of the faults that may appear.

With the information (x_{ij}, f_{ij}) that each agent a_{ij} has, the reconfiguration process is carried out, based on an agreement among agents. For this agreement, it is necessary that any information exchange is performed before the deadline of the task; nevertheless, this cannot be always achieved. The agreement demands network resources, which implies a raising of the traffic in the network, as well as possible saturation and data loss. So, the information available to achieve an agreement, and hence the reconfiguration, is incomplete [34]. To achieve reconfiguration of the task periods and the agreement among the agents, it is proposed to solve the data loss in the following way:

1. Modifying the periods for each sensor block. This attempts to avoid collisions and saturation in the network for each sampling period. The change of the periods of all elements in a distributed system is performed by modifying the ρ and λ parameters. This has been initially proposed by [29] and has been retaken here as reconfiguration strategy.
2. Data reconstruction. It is unavoidable that at a certain moment there is a large latency in data delivery, which generates a decrement in performance. Thus, it is proposed that the unavailable information can be reconstructed by means of a default value, previously defined. This commonly is obtained as an estimated value, given due to experience, or even a historical value, that could be the average of all previously registered values.

Agreement

Sensor redundancy increases measurement reliability. It is desired to have a high degree of security during data reading, so when a sensor presents any kind of fault,

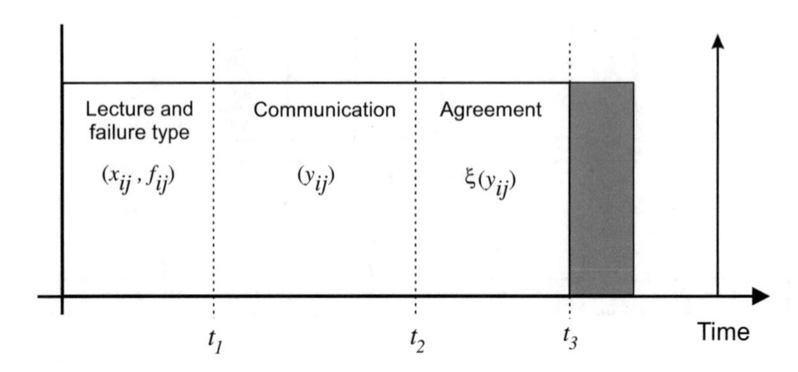

Fig. 3.13 An agent reaching an agreement value $\xi(y_{ij})$

a fault masking is activated, making use of a voting algorithm. Sensor information represents the dynamic state in the time t. During a fault scenario, local information that each sensor perceives is shared with other agents, to agree about a consolidated reading, and about performing reconfiguration. Measured signal information and fault classification (x_{ij}, f_{ij}) of each sensor agent are exchanged with its pairs within the same sensing block. Each agent counts then with a data in tuple $y_{ij} = (x_{i1}, f_{i1}), (x_{i2}, f_{i2}), \ldots, (x_{ij}, f_{ij})$ in which $i = 1, \ldots, m$ and $j = 1, 2, 3$. With this tuple values, each sensor agent performs an agreement $\xi(y_{ij})$, collaborating with other agents within the same sensor block. Each agent j within sensor block i provides a reading of the system state in time t_1: the measured signal x_{ij} and the fault classification f_{ij}. In time t_2, the agent has exchanged local information with the other agents in its sensor block. In time t_3, each agent has reached the agreement value $\xi(y_{ij})$ (Fig. 3.13).

Let us take the agreement $\xi(y_{ij})$ as a function that depends on the tuple of values y_{ij}. This function returns the agreed value of the measured consolidated signal $\bar{x}$, based on the classification of the fault, given in a sensor in time t_k. This agreed value prevails to be sent to the controller. Given m sensor blocks with triple redundancy, the agreement function $\xi(y_{ij})$ is proposed for $i = 1, 2, \ldots, m$ and $j = 1, 2, 3$ in the following way:

$$\xi = \begin{cases} \bar{x} = x_{i1} & f_{i1} = f_{i2} = f_{i3} \\ \bar{x} = \frac{(x_{i1}+x_{i2})}{2} & f_{i1} = f_{i2} \neq f_{i3} \\ \bar{x} = \frac{(x_{i1}+x_{i3})}{2} & f_{i1} = f_{i3} \neq f_{i2} \\ \bar{x} = \frac{(x_{i2}+x_{i3})}{2} & f_{i1} \neq f_{i2} = f_{i3} \\ \bar{x} = x_{i1} & f_{i1} \neq f_{i2} \neq f_{i3} \end{cases} \tag{3.19}$$

To study the impact of network utilisation on closed control loop, the 2-DOF Helicopter control model is built as a RTDS. Several nodes are connected through a common communication network. The experiments focus on network scheduling, and the main objective is to balance the amount of data sent through the network,

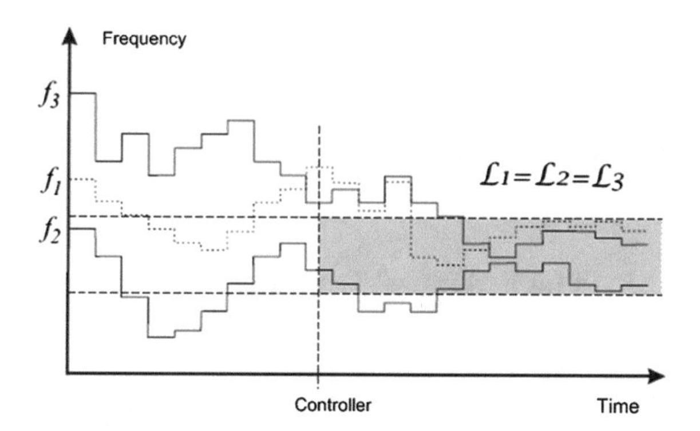

Fig. 3.14 A non-linear region L

in order to avoid latency and undersampling. Two network scheduling proposals are given to explore several aspects of control performance when the use of the network exceeds network bandwidth.

The performance criterion to quantify the system's quality performance uses the absolute value of the error IAE, as previously shown in Eq. 3.17.

The idea of bounded RTDS is implicit to manipulate the transmission frequencies and processing frequencies of periodic tasks. An approach to schedule a real-time distributed system based upon modifications on frequency transmission of individual components in the system is presented in [5]; this shows that scheduling of a distributed system can be accomplished through modifications on transmission frequencies into a region where the system performance is not affected. A linear time-invariant model in which the coefficients of the state matrix are the relations between the transmission frequencies of each node and through an LQR feedback controller to modify transmission frequencies bounded between maximum and minimum values of transmission. For instance, four measures are very important to implement a closed-loop control cycle in the helicopter system: θ, $\dot{\theta}$, ψ, $\dot{\psi}$. These are sensed by an equal number of nodes which in a distributed manner send sensed data θ_s, $\dot{\theta}_s$, ψ_s, $\dot{\psi}_s$ to the node controller. Each sensor performs a periodic task with a sampling period p_i.

This approach drives the frequency transmission through three parameters: minimum frequency f_m, real frequency f_r and maximum frequency f_x. The distributed system dynamics can be modelled as a linear time-invariant system whose state variables are transmission frequencies $[f_i = 1/p_i]$ of n nodes involved in it. There is a relationship between node frequencies and external input frequencies which serves as coefficients of the linear system; therefore, it is possible to control the NCS through input vector u such that outputs y are the node frequencies in a non-linear region L bounded by maximum and minimum transmission frequencies (Fig. 3.14).

The change in frequency transmission of each node through the time is represented using the relations of frequencies amongst a particular node and all nodes. It means that the frequency transmission rate of each node is influenced by changes in the transmission rate of other nodes.

This can be followed as presented in Sect. 3.3. Since nodes are considered autonomous its treatment should be different in order to stress the capability of decision taken in terms of this procedure. Therefore the use of DAI allows synchronising communication and process information.

3.5 The Effects of Time Delays

Consider a set of real-time tasks $\Gamma = \{\tau_1, \tau_2, \ldots, \tau_n\}$, where τ_i is the ith task; n is the number of tasks which are periodic, independent, and preemptive; c_i denotes the execution time of task τ_i; and p_i denotes the period of task τ_i. The approach here drives the Frequency Transmission (FT) based on three parameters: minimum frequency f_m, maximum frequency f_h, and real frequency f_r. FT dynamics can be modelled as a linear time-invariant subsystem, whose state variables are transmission frequencies of the sensor nodes involved on the system. Note that for each task of a sensor S_i, $i = 1, 2, \ldots, n$, frequency can be expressed as $f_i = 1/p_i$. Here, it is assumed that there is a relationship between frequencies, which define the internal dynamic of the network and the desired FT, and control input, which serve as coefficients of the linear system in Eqs. 3.7 and 3.8.

It is important to note that the relations between the frequencies of the n nodes leading to the system in Eqs. 3.7 and 3.8 are schedulable with respect to the network bandwidth. The relationship matrix is dependent on each case study, but it is mainly related to the fractions between each node transmission. Therefore, it is possible to control the FT through the input vector $u(k)$ such that the outputs $y(k)$ are in a region L that is non-linear, and where the system is schedulable. During the time evolution of the system, the output frequencies could be stabilised by a controller within the schedulability region L [23]. Figure 3.15 shows the dynamics of the frequency system and the desired effect when it is controlled. It also defines a common region L for a set of frequencies. Each task of the node i of the system starts with a frequency f_i, and the controller modifies the period p_i in order to converge in a region where the system performance is close to optimal.

The objective of modifying the frequency is to achieve coordination between the nodes to obtain the desired transmission frequencies.

Relationship Matrix. Let $a_{ij} \in A$ be given by a function of minimal frequencies f_m of node i, and $b_{ij} \in B$ be given by a function of maximal frequencies f_x:

$$a_{ij} = \phi\left(f_m^1, f_m^2, \ldots f_m^n\right) \tag{3.20}$$

$$b_{ij} = \gamma\left(f_x^1, f_x^2, \ldots f_x^n\right) \tag{3.21}$$

In this particular case, the matrix is built in terms of the local relations between the greatest common divisor, as shown in the next section; however, this is not the only procedure to be followed.

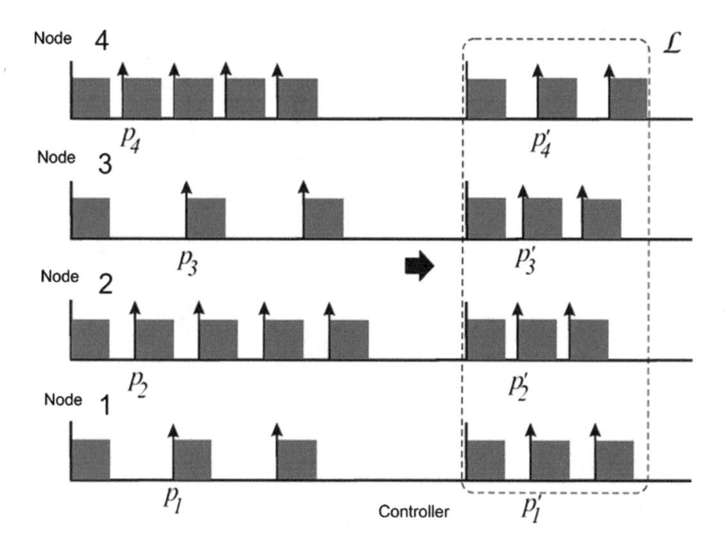

Fig. 3.15 Frequencies of transmission bounded by a schedulability region

The elements of the matrices for the system in Eqs. 3.20 and 3.21 are defined as follows:

$$a_{ij} = \begin{cases} \frac{\overline{\lambda}(f_m^1, f_m^2, \ldots, f_m^n)}{f_m^i} & i = j \\ \frac{f_m^j}{f_m^i} & i \neq j \end{cases} \tag{3.22}$$

$$b_{ij} = \begin{cases} f_x^i & i = j \\ 0 & i \neq j \end{cases} \tag{3.23}$$

$$c_{ij} = \begin{cases} 1 & i = j \\ 0 & i \neq j \end{cases} \tag{3.24}$$

$\overline{\lambda}(f_m^1, f_m^2, \ldots, f_m^n)$ is the greatest common divisor of the minimum frequencies. This is only written as $\overline{\lambda}$.

Control Design. The LQR controller is used to obtain the feedback matrix K. The input is given by a function of the minimal frequencies and the real frequencies of the n nodes in the distributed system, and hence:

$$u = K\left(\overline{f}_m - \overline{f}_r\right) \tag{3.25}$$

$$\overline{f}_m = \left[f_m^1, f_m^2, \ldots f_m^n\right]^T \tag{3.26}$$

$$\overline{f}_r = \left[f_r^1, f_r^2, \ldots f_r^n\right]^T \tag{3.27}$$

$$\overline{f}_d = \left[f_d^1, f_d^2, \ldots f_d^n\right]^T \tag{3.28}$$

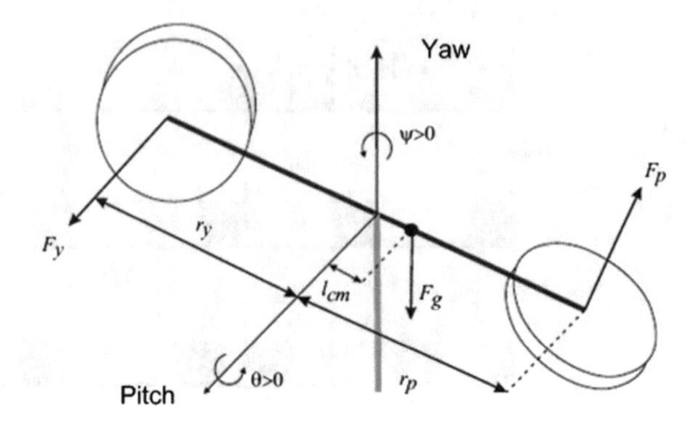

Fig. 3.16 Dynamics of the 2-DOF helicopter

Then the system in Eqs. 3.7 and 3.8 can be re-written as:

$$x(k+1) = (A - Bk^s)(\overline{f}_d - \overline{f}_r)(k) \tag{3.29}$$

where K is the control gain defined as the basics of a controller algorithm.

Helicopter Dynamic Model and its Control Design. The case study here is a prototype of a helicopter system, integrated to a CAN network with two propellers, driven by DC motors. The front propeller controls the elevation of the helicopter nose, about the pitch axis (θ), while the back propeller controls the side to side motions of the helicopter, about the yaw axis (ϕ). The pitch and yaw angles are measured using high-resolution encoders. A brief description of the helicopter model is presented. However, the detailed information can be found in [52]. The dynamics of the helicopter is developed based on kinetic and potential energy. This model is used to design a position controller. The helicopter centre of mass is described in xyz cartesian coordinates regarding the pitch and yaw angles (Fig. 3.16).

The Euler-Lagrange equations are used to obtain the non-linear motion equations for the 2 DOF Helicopter, which are used to derive the linear state model and, subsequently, to design the position controller. As the helicopter represents a non-linear system, it is required to perform a linearization around the point ($\theta_0 = 0$, $\psi_0 = 0$, $\dot{\theta}_0 = 0$, $\dot{\psi}_0 = 0$). From this, the linearization of the motion equation is obtained as follows:

$$\left(J_{eq,p} + m_{heli} l_{cm}^2\right) \dot{\theta} = K_{pp} V_{m,p} + K_{py} V_{m,y} - B_p \dot{\theta} - m_{heli} g l_{cm} \tag{3.30}$$

$$\left(J_{eq,y} + m_{heli} l_{cm}^2\right) \dot{\psi} = K_{pp} V_{m,y} + K_{py} V_{m,p} - B_p \dot{\psi} - 2 m_{heli} l_{cm}^2 \theta \psi \dot{\theta} \tag{3.31}$$

The state space is obtained following the linear model of in Eqs. 3.30 and 3.31, and solving for $x = [\theta, \psi, \dot{\theta}, \dot{\psi}]$.

$$\dot{x} = \begin{bmatrix} 0\,0 & 1 & 0 \\ 0\,0 & 0 & 1 \\ 0\,0 & -\frac{B_p}{J_{eq,p}+m_{heli}l_{cm}^2} & 0 \\ 0\,0 & 0 & -\frac{B_y}{J_{eq,y}+m_{heli}l_{cm}^2} \end{bmatrix} x + \begin{bmatrix} 0 & 0 \\ 0 & 0 \\ \frac{K_{pp}}{J_{eq,p}+m_{heli}l_{cm}^2} & \frac{K_{py}}{J_{eq,p}+m_{heli}l_{cm}^2} \\ \frac{K_{yp}}{J_{eq,y}+m_{heli}l_{cm}^2} & \frac{K_{yy}}{J_{eq,y}+m_{heli}l_{cm}^2} \end{bmatrix} \tag{3.32}$$

$$y = \begin{bmatrix} 1\,0\,0\,0 \\ 0\,1\,0\,0 \\ 0\,0\,1\,0 \\ 0\,0\,0\,1 \end{bmatrix} x \tag{3.33}$$

where Kpy, K_{yy}, K_{pp}, and K_{yp} are the torque-constants used to obtain coupled torques acting on the helicopter. $V_{m,p}$ is the input pitch motor voltage and $V_{m,y}$ is the input yaw motor voltage. For the state space model the input $u = \left[V_{m,p}, V_{m,y}\right]$ is the input vector, while $y = [x_1, x_2, x_3, x_4]$ is the output vector. Notice that, since the output matrix is the identity matrix, all states are measurable.

The model makes use of several Simulink™ and Matlab™ programs to develop the helicopter basic dynamics, by running a simulation of the closed-loop response, using the position controller. Regarding control issues, a LQR+I controller is designed. This controller regulates the pitch axis of the helicopter, using feed-forward (FF) and proportional-velocity (PV) compensator, while the and yaw axis only makes use of a PV control. The LQR+I controller uses an integrator in the feedback loop to reduce the steady-state error, by a feed-forward and proportional-integral-velocity (PIV) algorithms to regulate the pitch, and only a PIV to control the yaw angle. The LQR+I converges $\left(\theta, \psi, \dot{\theta}, \dot{\psi}\right) \rightarrow \left(\theta_d, \psi_d, \dot{\theta}_d, \dot{\psi}_d\right)$ where θ_d is the desired pitch angles and ψ_d is the desired yaw angle, such that:

$$\begin{bmatrix} u_p \\ u_y \end{bmatrix} = \begin{bmatrix} K_{ff} \frac{m_{heli} g l_{cm} \cos\theta_d}{K_{pp}} \\ 0 \end{bmatrix} \tag{3.34}$$

The addition of an integrator requires to introduce the states $\dot{x}_5 = \theta$ and $\dot{x}_6 = \psi$, so the linear state-space model is augmented as:

$$\dot{x} = \begin{bmatrix} 0\,0 & 1 & 0 & 0\,0 \\ 0\,0 & 0 & 1 & 0\,0 \\ 0\,0 & -\frac{B_p}{J_{eq,p}+m_{heli}l_{cm}^2} & 0 & 0\,0 \\ 0\,0 & 0 & -\frac{B_y}{J_{eq,y}+m_{heli}l_{cm}^2} & 0\,0 \\ 1\,0 & 0 & 0 & 0\,0 \\ 0\,1 & 0 & 0 & 0\,0 \end{bmatrix} x + \begin{bmatrix} 0 & 0 \\ 0 & 0 \\ \frac{K_{pp}}{J_{eq,p}+m_{heli}l_{cm}^2} & \frac{K_{py}}{J_{eq,p}+m_{heli}l_{cm}^2} \\ \frac{K_{yp}}{J_{eq,y}+m_{heli}l_{cm}^2} & \frac{K_{yy}}{J_{eq,y}+m_{heli}l_{cm}^2} \\ 0 & 0 \\ 0 & 0 \end{bmatrix} u \tag{3.35}$$

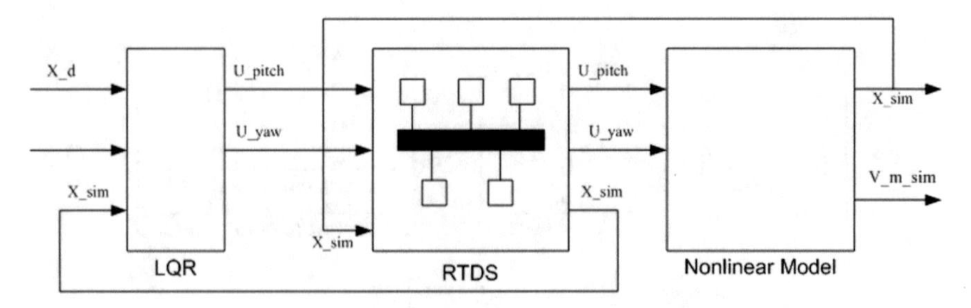

Fig. 3.17 Block representation of a NCS of eight nodes with a real-time kernel, connected through a network type CSM/AMP (CAN)

$$y = \begin{bmatrix} 1&0&0&0&0&0 \\ 0&1&0&0&0&0 \\ 0&0&1&0&0&0 \\ 0&0&0&1&0&0 \\ 0&0&0&0&1&0 \\ 0&0&0&0&0&1 \end{bmatrix} x \tag{3.36}$$

Using the adequate Q and R weighting matrices, the control gain is as follows:

$$k = \begin{bmatrix} 18.9 & 1.98 & 7.48 & 1.53 & 7.03 & 0.77 \\ -2.22 & 19.4 & -0.45 & 11.9 & -0.77 & 7.03 \end{bmatrix} \tag{3.37}$$

Thus, the LQR+I controller is:

$$\begin{bmatrix} u_p \\ u_y \end{bmatrix} = \begin{bmatrix} K_{ff} \frac{m_{hrli} g l_{cm} \cos\theta_d}{K_{pp}} \\ 0 \end{bmatrix} - \begin{bmatrix} k_{11} & k_{12} & k_{13} & k_{14} \\ k_{21} & k_{22} & k_{23} & k_{24} \end{bmatrix} \begin{bmatrix} \theta - \theta_d \\ \psi - \psi_d \\ \dot{\theta} \\ \dot{\psi} \end{bmatrix} - \begin{bmatrix} \int k_{15} (\theta - \theta_d) + \int k_{16} (\psi - \psi_d) \\ \int k_{25} (\theta - \theta_d) + \int k_{26} (\psi - \psi_d) \end{bmatrix} \tag{3.38}$$

In order to study the impact of network utilisation on closed control loop, the 2-DOF Helicopter control model is built as a RTDS. Several nodes are connected through a common communication network. The experiment focuses on network scheduling and the main objective is to balance the amount of data sent through the network, in order to avoid latency and undersampling. Let us consider the special case of a RTDS, in which the structure is based on sensor, controller, actuator, and master scheduler nodes. Figure 3.17 shows the NCS consisting of eight nodes with a real-time kernel, connected through a network type CSM/ AMP (CAN). The rate of sending data is 10 Mb/s, which is not likely for data loss. The real-time simulation tool is Truetime [53], which is based on Matlab™/Simulink™.

Four sensor nodes execute periodic tasks to sense control signals, as well as other additional periodic tasks. Each task has a period p_i and time consumption c_i as shown in Fig. 3.17. The sensed control signals are $x = (\theta, \psi, \dot{\theta}, \dot{\psi})$. This model has a controller node, depicted on the left side. This controller takes the control law from the LQR module using a task, which activates by event. The time consumption of the controller task is the maximum average time it takes to compute the control law. The controller node uses the values from sensors, and sends control outputs u_p and u_y, that correspond to the pitch and yaw voltages. Two actuator nodes, located on the bottom right corner shown in Fig. 3.17, receive signals from the controller node. Finally, the scheduler node, located on the top right corner, organises the activity of the other seven nodes, and it is responsible for periodic allocation bandwidth, used by these nodes. Each node initializes specifying the number of inputs and outputs of the respective True Time kernel block, defining a scheduling policy, and creating periodic tasks for the simulation. These tasks involve parameters about the periodic times and the consumption times. The task periodic times define the time interval between tasks, whereas the consumption times refer to the execution time of the task. The previous system has been included in a feedback control loop of 2-DOF helicopter model.

Changes in the real-time task parameters of the RTDS commonly impact on network utilisation, and therefore, on the control performance [13, 32]. The problem to tackle, thus, is to find a proper way to schedule the common communication network of the RTDS, based on managing an accurate sampling period, and capable of keeping both, the network load and required integrated performance.

Previous work is related to the scheduling transmission rate of data in a real-time distributed systems based on the modification of the sampling periods and develops the frequency transition model, as described in Sect. 2.3. Reference [29] design a global scheduling strategy based on the analysis of NCS. They show that the performance of the system depends not only on the sampling periods of its individual components but also on the time dispersions amongst these periods. The scheduling strategy consists of a global operating period for the whole system called base period. In addition, there is a common periodic range for every node participating in the NCS. This period range is obtained by moving away from the base period within a certain percentage, called dispersion factor. Base period and dispersion factor are used to obtain the actual operational period for each node i of the system: $n_i = \rho(1 \pm \lambda)$. Some tests are presented in [29] to quantify the NCS quality performance under a particular scheduling strategy, and to evaluate the performance of a NCS. When all nodes are racing to obtain the network and get bandwidth, according to the network access control algorithm, the system tends to be unstable. As part of the scheduling strategy, the scheduler node allocates a bandwidth share to every node, by means of assigning a time-window to transmit, independently from the network protocol used. This has to be solely considered as the network access controller. To have a performance criterion, and thus, quantify the system's quality performance, the integral of the absolute value of the error IAE is used (Eq. 3.17).

The 2-DOF Helicopter system presents a stable behaviour using a base period in the interval [0.005, 0.017] seconds and a fixed dispersion of 5% for all nodes; however when the task period is out of this interval and/or the scheduler does not assign a proper bandwidth, the system becomes easily unstable (Fig. 3.18).

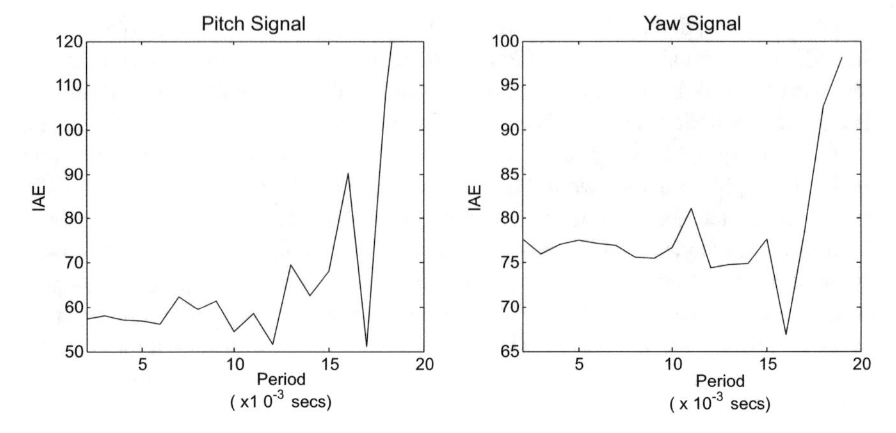

Fig. 3.18 The system becomes unstable

It is important to emphasise that a successful network management is a key point to achieve system schedulability, and thus, obtain a good performance. A disadvantage of the previous proposal is due to the static scheduling model, which means that accurate real-time parameters are computed off-line, and there is no mechanism to modify these parameters on-line. This would be very useful under fault scenarios. Here, the focus is on the sensor nodes, with the objective of controlling the data FT through the system in Eqs. 3.7 and 3.8.

Defining $e(k) = (f_d - f_r)(k)$, Eq. 3.29 becomes:

$$e(k+1) = (A - Bk^s)e(k) \tag{3.39}$$

Now defining $e_5(k) = \varepsilon_d(k) - \varepsilon_r(k)$ as the augmented state, the system is as follows:

$$\begin{bmatrix} e_1^{k+1} \\ e_2^{k+1} \\ e_3^{k+1} \\ e_4^{k+1} \\ e_5^{k+1} \end{bmatrix} = \begin{bmatrix} \frac{\bar{\lambda}}{f_m^1} - f_x^1 k_1^s & \frac{f_m^2}{f_m^1} & \frac{f_m^3}{f_m^1} & \frac{f_m^4}{f_m^1} & 0 \\ \frac{f_m^1}{f_m^2} & \frac{\bar{\lambda}}{f_m^2} - f_x^2 k_2^s & \frac{f_m^3}{f_m^2} & \frac{f_m^4}{f_m^2} & 0 \\ \frac{f_m^1}{f_m^3} & \frac{f_m^2}{f_m^3} & \frac{\bar{\lambda}}{f_m^3} - f_x^3 k_3^s & \frac{f_m^4}{f_m^3} & 0 \\ \frac{f_m^1}{f_m^4} & \frac{f_m^2}{f_m^4} & \frac{f_m^3}{f_m^4} & \frac{\bar{\lambda}}{f_m^4} - f_x^4 k_4^s & 0 \\ c_1 & c_2 & c_3 & c_4 & 1 - k_5^s \end{bmatrix} \begin{bmatrix} e_1^k \\ e_2^k \\ e_3^k \\ e_4^k \\ e_5^k \end{bmatrix} \tag{3.40}$$

where c_1, c_2, c_3, c_4 are the consumption times by each process, considering the periodic tasks within their respective periods; and k_i^s represents the control gains. Thus, the output vector is:

$$\begin{bmatrix} y_1^{k+1} \\ y_2^{k+1} \\ y_3^{k+1} \\ y_4^{k+1} \\ y_5^{k+1} \end{bmatrix} = \begin{bmatrix} 1\,0\,0\,0\,0 \\ 0\,1\,0\,0\,0 \\ 0\,0\,1\,0\,0 \\ 0\,0\,0\,1\,0 \\ 0\,0\,0\,0\,1 \end{bmatrix} \begin{bmatrix} f_r^1 \\ f_r^2 \\ f_r^3 \\ f_r^4 \\ \varepsilon_r \end{bmatrix} \tag{3.41}$$

Since the FT data of the network nodes using a NCS can increase without limit due to uncertainties during sensing, the system may become unstable; similarly, a low FT data results in undersampling, which leads to control performance weakness. Therefore, it is important to manage the sensor nodes dynamically, to ensure that they are restricted to the schedulability region, where the system is stable. It is possible to perform a network planning outline; however, uncertainties inherent to the network, such as traffic or time delays, require dynamic management of the transmission frequencies. Although in practice this control accepts frequencies outside of the stability region, it is possible to balance the use of the network and maintain the system at acceptable performance levels.

3.6 Concluding Remarks

This chapter provides a review of distributed systems modelling, considering several approximations, such as real-time strategies, coordinated task strategies, consensus approaches, as well as other with major importance in terms of time delays representation.

To achieve this approximation, it is necessary to define several characteristics of tasks representation, like preemption, bounded time, synchronisation, as well as certain kind of identification. This representation and bounded relations possible to determine the effects of time delays from a scaling perspective. For instance, the idea of scale is well reviewed in terms of local and global approximation, where the effects of local time delays may have a global repercussion since there is an incremental effect among local effects on global behaviour.

In this chapter, a special interest is given in terms of local scheduling strategies, as well as computer network control communication, following several validated position points. Moreover, the review of coordinated tasks proposes an affordable situation in order to enhance a distributed system through consensus among elements, either global or semi-global effects. Finally, in any case, the focusing effect is to determine the time delays into a dynamic system, regardless of the scale used.

References

1. Heemels, W.P.M.H., Siahaan, H., Juloski, A., Weiland, S.: Control of quantized linear systems: an optimal control approach. In: Proceedings of American Control Conference, Denver, CO, pp. 3502–3507 (2003)
2. Luck, R., Ray, A.: An observer-based compensator for distributed delays. Automatica **26–5**, 903–908 (1990)
3. Coulouris, G., Dollimore, J., Kindberg, T., Blair, G.: Distributed Systems. Addison Wesley (2012)
4. Tanenbaum, A., Van-Steen, M.: Distributed Systems: Principles and Paradigms. Prentice Hall (2007)
5. Esquivel-Flores, O.A.: Estudio de Sistemas reconfigurables en tiempo real con base en un sistema multiagente reactivo. Posgrado en Ciencias e Ingenieria de la Computacion, UNAM. 23 Enero (2013)
6. Association, Open DeviceNet Vendors: DeviceNet Specification, 2.0 (1997)
7. Liu, F., Narayanan, A., Bai, Q.: Real-Time Systems (2000)
8. Hadzilacos, V., Toueg, S.: A Modular Approach to the Specification and Implementation of Fault-Tolerant Broadcasts. Technical Inform, Department of Computer Science, Cornell University, Ithaca N.Y (1994)
9. Lamport, L.: Time clocks and the ordering of the events on distributed systems. Commun. ACM **21**(7) (1978)
10. Khanna, V.K., Singh, S.: An improved pigback ethernet protocol and its analysis. Comput. Netw. ISDN Syst. **26**(11), 1437–1446 (1994)
11. Wheelis, J.D.: Process control communication: token bus, CSMA/CD, or token ring. ISA Trans. **32**(2), 193–198 (1993)
12. Koubias, S.A., Papadopoulos, G.D.: Modern fieldbus communication architectures for real-time industrial applications. Comput. Ind. **26**(3), 243–252 (1995)
13. Lian, F.L., Moyne, J.R., Tilbury, D.M.: Performance evaluation of control networks: Ethernet, ControlNet, and DeviceNet. IEEE Control Syst. Mag. **21**, 66–83 (2001)
14. Tanenbaum, A.S.: Computer Networks, 3rd edn. Prentice-Hall Inc. (1996)
15. Verriest, E., Egerstedt, M.: Control with delayed and limited information: a first look. In: Proceedings of the 41st IEEE Conference on Decision and Control, Las Vegas, USA, pp. 1231–1236 (2002)
16. Edison, J., Cole, W.: Ethernet rules closed-loop systems. InTech, 39–42 (1998)
17. Kaplan, G.: Ethernets winning ways. IEEE Spectr. **38**(1), 113–114 (2001)
18. Imer, O.C., Yuksel, S., Basar, T.: Optimal control of LTI systems over unreliable communication links. Automatica **42**(9), 1429–1439 (2006)
19. Walsh, G.C., Beldiman, O., Bushnell, L.: Error encoding algorithms for networked control systems. In: Proceedings of the 38th IEEE Conference on Decision and Control, vol. 5, pp. 4933–4938 (1999)
20. Tanenbaum, A.: Distributed Operating Systems. Prentice Hall (1995)
21. Xia, F., Sun, Y.: Control and Scheduling Codesign: Flexible Resource Management in Real-Time Control Systems. Springer (2008)
22. Kumar, S., Raghavan, V., Deng, J.: Medium access control portocols for ad hoc wireless networks: a survey. Ad Hoc Netw. **4**, 326–358 (2006)
23. Esquivel-Flores, O., Benitez-Perez, H., Mendez, E., Menendez, A.: Frequency transition for scheduling management using dynamic system approximation for a kind of NCS. ICIC Express Lett. part B Appl. **1**(1), 93–98 (2010)
24. Benitez-Perez, H., Garcia-Nocetti, F.: Reconfigurable Distributed Control. Springer, Berlin, Germany (2005)
25. Quanser: 2-DOF Helicopter, User and Control Manual. Quanser, Speciality Experiments: 2-DOF Helicopter (2007)

26. Cervin, A., Ohlin, M., Henriksson, D.: Simulation of networked control systems using truetime. In: Proceedings of the 3rd International Workshop on Networked Control Systems: Tolerant to Faults, Nancy, France (2007)
27. Cervin, A., Henriksson, D., Lincoln, B., Eker, J., Arzen, K.-E.: How does control timing affect performance? Analysis and simulation of timing using jitterbug and truetime. IEEE Control Syst. **23**(3), 16–30 (2003)
28. Mendez-Monroy, P.E., Benitez-Perez, H.: Supervisory fuzzy control for networked control systems. Int. J. Innov. Comput. Inf. Control Express Lett. **3**(2), 233–240 (2009)
29. Menendez-Leonel, A., Benitez-Perez, H.: Scheduling strategy for real-time distributed systems. J. Appl. Res. Technol. **8**(2), 177–185 (2010)
30. Liu, C., Layland, J.: Scheduling algorithms for multiprogramming in a hard-real-time environment. J. Assoc. Comput. Mach. **20**(1), 46–61 (1973)
31. Balbastre, P.: Modelos de Tareas para la Integracion del Control y la Planificacion en Sistemas de Tiempo Real. Universidad Politecnica de Valencia, Tesis Doctoral (2002)
32. Lian, F., Moyne, J., Otanez, P., Tilbury, D., Moyne, J.: Design of sampling and transmission rates for achieving control and communication performance in networked multi-agent system In: Proceedings of American Control Conference, Denver, USA, 4–6 June, pp. 3329–3334 (2003)
33. Weiss, G.: Multiagent Systems a Modern Approach to Distributed Artificial Intelligence. MIT (1999)
34. Lynch, N.: Distributed Algorithms. Morgan Kauffman (1996)
35. Bond, A., Gasser, L.: Readings in Distributed Artificial Intelligence. Morgan Kaufmann (1988)
36. Huhns, M.N., Singh, M.P. (eds.): Readings in Agents. Morgan Kaufmann (1998)
37. Benitez-Perez, H., Garcia-Zavala, A., Garcia-Nocetti, F.: Aproposal for online reconfiguration based upon a modification of planning scheduling and fuzzy logic contol law response. In: ISSADS. Lecture Notes Computer Science, pp. 141–152. Springer (2005)
38. Jennings, N.: Cooperation in Industrial Multiagent Systems. World Scientific (1994)
39. Wooldridge, M., Jennings, N.: Intelligent Agents. Lecture Notes in Artificial Intelligence, N. R. (1995)
40. Cenjor, A., Garcia, A.: Control basado en agentes mejorados con tecnologia auto-id. RIAII **2**(3), 48–60 (2005)
41. Higuera, A.G., Montalvo, A.C.: Sistema heterarquico de control basado en agentes para sistemas de fabricacion: la nueva metodologia proha. Revista Iberoamericana de Automatica e Inform. Ind. RIAI **4**(1), 83–94 (2007)
42. Cena, C.G., Saltaren, R., Blazquez, J.L., Aracil, R.: Desarrollo de una interfaz de usuario para el sistema robotico multiagente smart. Revista Iberoamericana de Automatica e Inform. Ind. RIAI **7**(4), 17–27 (2010)
43. Ramirez-Gonzalez, T., Quinones-Reyes, P., Benitez-Perez, H., Laureano-Cruces, A., Garcia-Nocetti, F.: Reconfigurable fuzzy Takagi Sugeno networked control using cooperative agents and local fault diagnosis. In: IEEE International Symposium on Intelligent Signal Processing, 2007, WISP 2007, pp. 1–5. IEEE (2007)
44. Corkill, P.: Collaborating software: blackboard and multiagent systems and the future. In: Proceedings on the International Lisp Conference, vol. 3, pp. 123–138 (2003)
45. Ramirez, T.: Algoritmo de Planificacion en un Sistema de Control Distribuido basado en una Arquitectura Multiagente en Tiempo Real. Universidad Autonoma Metropolitana, Tesis de Maestria (2006)
46. Oosterom, M.: Soft computing methods in flight control systems design. Ph.D. University Technology Delft (2005)
47. Benitez-Perez, H., Garcia-Nocetti, F.: Reconfigurable distributed control using smart peripheral elements. Control Eng. Pract. **11**(9), 975–988 (2003)
48. Lunze, J., Steffen, T.: Hybrid reconfigurable control. In: Modelling, Analysis, and Design of Hybrid Systems, pp. 267–284. Springer, Berlin, Heidelberg (2002)

49. Lunze, J., Rowe-Serrano, D., Steffen, T.: Control reconfiguration demonstrated at a two-degrees-of-freedom helicopter model. In: Proceedings European Control Conference, Cambridge, UK (2003)
50. Latif-Shabgahi, G., Bennett, S., Bass, J.M.: Smoothing voter: a novel voting algorithm for handling multiple errors in fault-tolerant control systems. Microprocess. Microsyst. **27**(7), 303–313 (2003)
51. Latif-Shabgahi, G.R.: A novel algorithm for weighted average voting used in fault tolerant computing systems. Microprocess. Microsyst. **28**(7), 357–361 (2004)
52. Quanser: Quanser 2 DOF Helicopter, User and Control Manual, Quanser Innovative-Educate (2006)
53. Ohlin, M. Henriksson, D., Cervin, A.: True Time 1.5 Reference Manual. Department of Automatic Control, Lund University (2007)

Chapter 4
Design of Networked Control System

Abstract This chapter presents the control strategies and codesign based on the NCS models presented in Chap. 2. Three methodologies are proposed focusing on vanishing the perturbations generated by the network, such as time delays larger than a sampling period and lost packets. First, an adaptive fuzzy control is developed according to the scheduling algorithm and the known and bounded time delay, the stategy us a fuzzy model and a LQR control design to modify the control input according to the time delay. A sampling frequency control is presented where the transmission frequiencies are modified into a region according the quality of services into the network. Finally, a codesign strategy is reviewed where the quality of service and the quality of control are trade-off with two fuzzy model, one fuzzy model modifies the input control based on the current sampling period and other fuzzy model modifies the next sampling period based on the time delays and lost packets in a time-lapse.

4.1 Classical Approaches

A feedback control system, in which control loops are closed via a communication network, is called Networked Control System (NCS). Thus, sensors, actuators, controllers and others (monitors, etc.) are interconnected via one or several communication networks. The main advantages of this kind of systems are (a) their low cost, (b) the small volume of wiring, (c) the distributed processing, (d) their simple installation, (e) their maintenance, and (f) their reliability.

Recently, much attention has been paid to control design and stability analysis of NCSs [1–3]. The key problems to solve are minimized the effect of network-induced time delays and packet loss that degrade system performance. In general, time delay has been considered as constant, time varying, or even random. Time delay, as well as packet loss, depends on several factors such as the scheduler, network type, architecture, operating systems, processor conditions, etc. [4] as reviewed in Chap. 3. When the time delay is less than the sampling period of NCS, results indicate that time delay degrades the system performance. However, this situation is possible to be solved [5]. When the time delay is greater than sampling period, with a varying

H. Benítez-Pérez et al., *Control Strategies and Co-Design of Networked Control Systems*, Modeling and Optimization in Science and Technologies 13,
https://doi.org/10.1007/978-3-319-97044-8_4

or random behaviour, the performance of a NCS is considerably reduced. Hence, it is necessary to analyse time delays and packet loss to develop an efficient approach for reducing its effect on NCS. For instance, Nilsson analyses important facets of NCSs [4], introducing models for the time delays of a NCS. First, time delays are modelled as fixed values; later, they are modelled as independently random values; and finally, they developed a Markov process. Nilsson introduces optimal stochastic control theorems for NCSs, based on independently random and Markovian delay models, and considering time delays smaller than a sampling period.

Another approach is the well known Takagi-Sugeno-Kang (TSK) fuzzy models, which are qualified to represent a certain class of non-linear dynamic systems [5], and many control techniques have been developed following this approach. In particular, regarding NCS designs that make use of the TSK fuzzy models, some results have recently been published [6, 7]. In [6], a control with fault detection for NCSs with Markov delays is addressed, in which a linear plant is modelled in the discrete-time domain. A set of TSK fuzzy rules is used to deal with network-induced delays. Further, results in [6, 8] are formulated in the continuous time domain, where the TSK fuzzy systems with norm-bounded uncertainties are used to characterise the non-linear NCSs. Nilsson [4] proposes a LQR (Linear Quadratic Regulator) based on Fuzzy Logic. It presents a method for online estimation of the time delays for an invariant sampling period, and it is assumed that delays being less than a sampling period. Takaba [9] shows a robust control design for NCS polytopic systems, with time delays uncertain with PID controllers, but it only presents numerical examples attempting to show the effectiveness of the method.

Regarding the topic of time delays, [10] presents the effect of modelling time delays with a variable sampling period in discrete time systems. A similar approximation has been presented in [9], where a specific study on time delays is reviewed, as well as an application strategy for NCS. Walsh et al. [11] shows a practical approximation of time delays, assuming a regular computer network.

All these previous works provide some control or scheduling strategy, aiming for robustness to constant, and even varying, time delays. Similarly, for packet loss, several approaches assume a linear and available system model, which is periodically sampled. This means that some restrictions have to be taken into consideration for the implementation of control systems. Thus, here it is proposed a controller to stabilise non-linear systems, assuming that there is no model of the system-network. The only knowledge available is a continuous linear model, is used fuzzy controllers and assuming statistical values of RTT delay and packet loss due to the network activity as presented in [12].

NCSs have been focused on how time delays modify the dynamics of the system's behaviour during time response. Classical approaches have been reviewed in [13]. They suppose beforehand that time delays are mainly static and known. In these cases, when time delays present a dynamic behaviour, the dynamic modelling becomes quite complex, and in terms of the non-linearities presented. Therefore, it is necessary to build a mathematical approximation, capable of tackling the dynamic behaviour of time delays. One approach has been modelling the stochastic behaviour of time delays, as explored in [13], where the behaviour of time delays is modelled in terms of a Brownian system. An alternative approximation makes use of queueing theory, describing system performance in terms of a stochastic procedure for time delays. This approach tends to be appealing, particularly when non-linear computer networks are used, such as Ethernet, in large scale networks, in which several routers are involved for communication [13].

Another approach has been to consider real-time systems in which strictly bounded time is the main goal. Therefore, time delays are known, even during dynamic conditions. Classic approximations are presented by [5]. In this, the results are valid, although restricted by certain conditions like bounded time delays. In this case, the use of scheduling algorithms is explored, with interesting results on system stability.

An interesting approach proposes to control the frequency transmission from the communication channels, to guarantee how time delays behave in short and long terms. In this case [14], they propose a classical approach, using LQG, to guarantee time response from communication channels. However, this is only a first attempt over a complex problem, in which several conditions still have to be explored to overcome the natural restrictions from a LQG approximation. In terms of stability, a strong condition is reviewed by [2]. In this, a Lyapunov Krasovskii strategy is presented as a strong mathematical tool to determine when stochastic varying time delays take place. Time delays tend to be bound to certain realistic conditions, in terms of the behaviour of the computer network. Another approach makes use of the TSK Fuzzy control [15], in which there is a strong gain scheduling technique that combines several linear scenarios into a global stable system. In this, the use of linear matrix inequalities is a mechanism to design and ensure a global stable situation. One important condition is the restriction of the system in terms of observability for global purposes. A holistic review of this case has been studied in [16], in which several time delay conditions are determined, such as oversampling, package loss, long time delays, and stochastically bounded time delays. This work provides a clear approximation to time delays in several local conditions, with the compromise to guarantee global stability.

A final approximation for NCS is based on the synchronisation among nodes belonging to a complex system. Several approximations have been developed to the goal of fixed time delays, like extended states for synchronisation purposes of LMI procedures.

4.2 Adaptive Fuzzy Control

In this section, the objective is to present a adaptive control strategy developed from the time delay knowledge, and following scheduling approximation, where time delays are known and bounded, according to the scheduling algorithm used. The proposed scheduling strategy here has been reviewed in [17, 18]. It pursues to tackle local faults in terms of fault tolerance. In this situation, uncertain time delays would be inevitable.

The case study here makes use of the classical Earliest Deadline First (EDF) and the Priority Exchange (PE) algorithm to decompose timelines, and the respective inherent time delays. For instance, time delays are supervised for a number of tasks as follows:

$$\varepsilon_1 \rightarrow \varepsilon_n / T_1 \rightarrow T_n \tag{4.1}$$

Priority is provided by the EDF algorithm, in which the process with the closest deadline has the highest priority. However, when an aperiodic task appears, it is necessary to deploy other algorithms to cope with the current conditions. To do so, the PE algorithm is used to manage the spare time from the EDF algorithm. The PE algorithm [18] uses a virtual server that deploys a periodic task with the highest priority, to provide enough computing resources for aperiodic tasks. This simple procedure provides an approximate, deterministic, and dynamic behaviour, within the group of included processes. In this case, time delays can be deterministic and bounded.

Example 4.1 Consider the group of tasks as shown in Table 4.1. In this case, consumption times, as well as periods, are given in terms of integer units. Remember: the server task is the time given for an aperiodic task to take place on the system.

The result of the ordering, based on PE, is presented in Fig. 4.1.

Based on this dynamic scheduling algorithm, time delays are given as current calculations in terms of task ordering. In this case, every time that the scheduling algorithm takes place, the global time delays are modified in the short and long term.

For instance, consider the following example, in which four tasks are set, and two aperiodic tasks take place at different times, giving different events with different time delays (Table 4.2).

The following task ordering is shown in Fig. 4.2, using the PE algorithm, in which clearly time delays appear.

Now, from this, a resulting ordering of these tiny time delays are given for two scenarios, as shown in Fig. 4.3.

These two scenarios present two different local time delays that need to be taken into account beforehand to settle the related time delays, according to scheduling approach and control design. These time delays can be expressed in terms of local relations between both dynamical systems. These relations are the actual and possible delays, bounded as marked limits of possible and current scenarios. Then, delays may be expressed as local summations with a high degree of certainty.

Table 4.1 First example for PE algorithm

Name	Consumption (in units)	Period (in units)
Task 1	1	5
Task 2	2	9
Task 3	1	8
Server	1	11

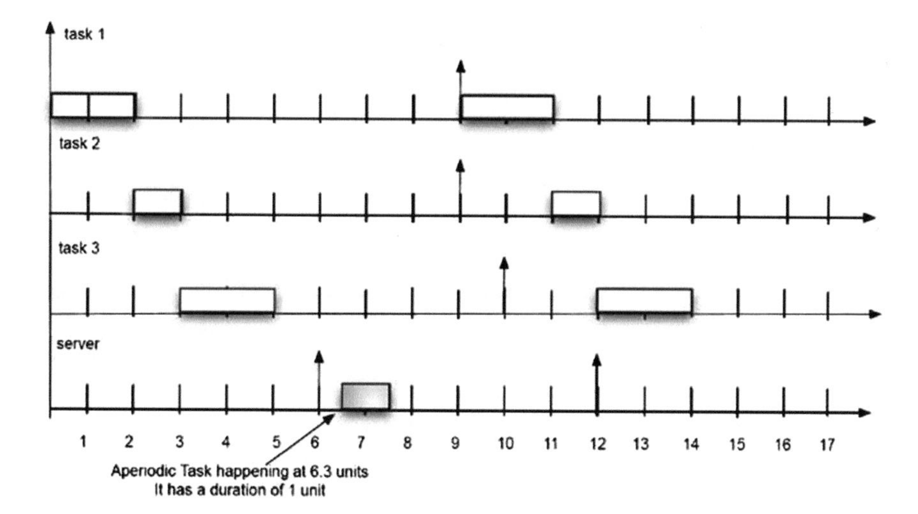

Fig. 4.1 Related organization for PE of tasks in Table 4.1

Table 4.2 Second example of PE

Name	Consumption (in units)	Period (in units)
Task 1	1	5
Task 2	2	9
Task 3	1	8
Server	1	11
Aperiodic task1 (ap1)	0.9	It occurs at 9
Aperiodic task (ap2)	1.0	It occurs at 13

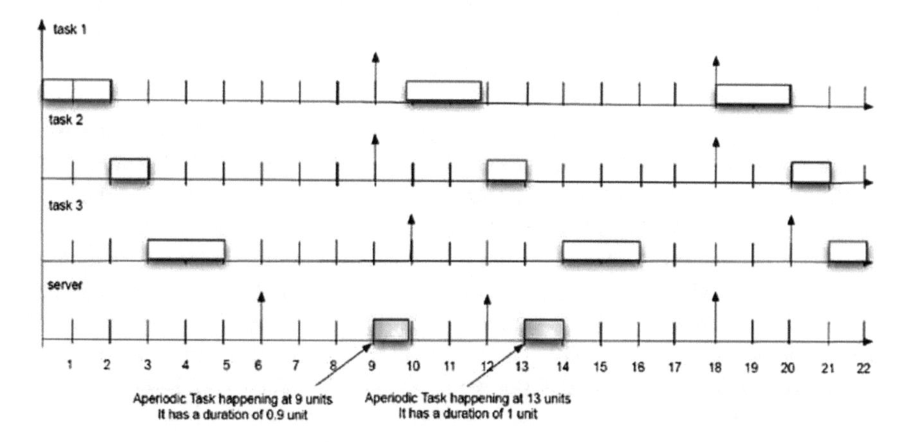

Fig. 4.2 Task organizations considering the second example for the PE algorithm

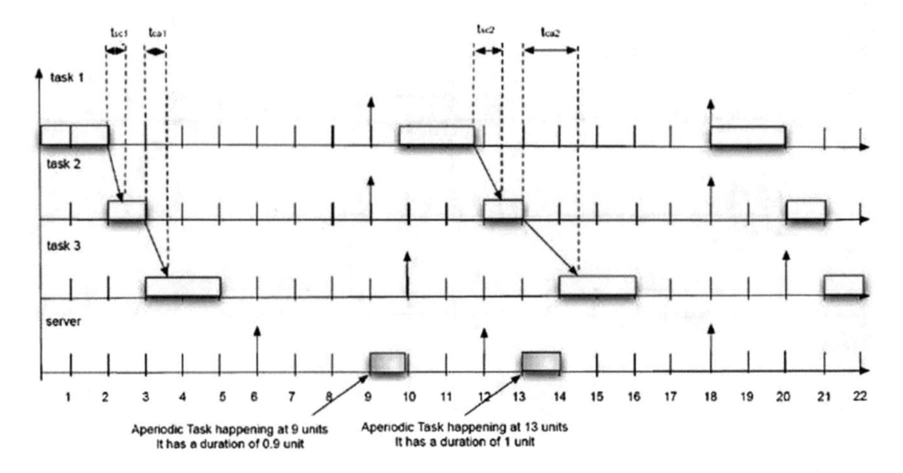

Fig. 4.3 Related time delays are depicted according to both scenarios

In the last example, during the second scenario, a total delay is given as:

$$\begin{aligned} Total Delay = {} & consumption\,Time\,Delay\,Aperiodic\,Task1 \\ & + consumption\,Time\,Delay\,Task1 \\ & + tsc2 + consumption\,Time\,Delay\,Task2 \\ & + consumption\,Time\,Delay\,Aperiodic\,Task2 \\ & + consumption\,TimeDelay\,Task3 \end{aligned} \tag{4.2}$$

Now, from this example, $lp = 2$ $(consumption\,Time\,Delay\,Aperiodic\,Task1 + consumption\,Time\,Delay\,Task1 + tsc2 + consumption\,Time\,Delay\,Task2)$ and $lc = 3$ $(consumption\,Time\,Delay\,Aperiodic\,Task2 + consumption\,Time\,Delay\,Task3)$. lp and lc are the total number of local delays within one scenario, from sensor to control, and from control to actuator, respectively. lp and lc are used as limits of current delays per scenario, as shown in Eqs. 2.10 and 4.4. These limits are modified according to the allowed scenario that guarantees schedulability during online process. Nevertheless, this execution need to be bounded by the only considered scenarios from TSK modelling as presented in previous chapter.

Having defined time delays as result of a scheduling approximation, several scenarios are potentially present following this time delay behaviour. In fact, the number of scenarios is finite, since the combinatorial formation is bounded. Therefore, any strategy in order to design a control law needs to take into account gain scheduling approximation, and thus, retaking the fuzzy model in Sect. 2.2, the controller is defined as follows:

$$x_c(k+1) = \sum_{j=1}^{r} h_j(A_j^c x_c(k) + \rho_j^c B_j^c u_c(k - \tau_j^{sc})) \tag{4.3}$$

$$y_c(k) = \sum_{j=1}^{r} h_j C_j^c x_c(k)$$

where the indexes c correspond to the dynamics of the controller.

The time delay representation in discrete system in the range $[t_i, t_{i+1}]$ is:

$$B_j^c = \sum_{j=1}^{l_c} \int_{t_i}^{t_{i+1}} e^{A_j^c t} \hat{B}_j^c dt \tag{4.4}$$

where $\hat{B}_j^c$ are the sets of input matrices of the continuous time models of the controller. According to (2.5) l_c represent the total number of local time delays that appears per scenario, and they are the source of τ_j^{sc}. The output is gathering as following:

$$y_c(k) = \sum_{j=1}^{r} h_j C_j^c x_c(k) \rightarrow u_p\left(k - \tau_j^{ca}\right) = y_c(k - \tau_j^{ca}) \tag{4.5}$$

A complete representation of process and control is obtained in terms of a system in equilibrium, given as:

$$x_c(k+1) = \sum_{j=1}^{r} h_j \left(A_j^c x_c(k) + \rho_j^c B_j^c \sum_{i=1}^{r} h_i C_i^c x_p(k - \tau_i^{sc}) \right) \tag{4.6}$$

From the last equation, the related control dynamics are expressed as A_j^c, B_j^c, and C_j^c, where j is the index with respect to each scenario.

With the fuzzy model in Eq. 2.12, the estimated state of system is obtained by compensating the time delays and variable sampling intervals. The action is to smoothly switch between discrete models, to generate the best estimate of state, according to the estimated time delay $\hat{\tau}(k)$. Thus, let us propose a design for the fuzzy controller, using a fuzzy feedback control law like:

$$u(k) = -\sum_{j=1}^{r} h_j \bar{L}_j x(k) \qquad j = 1, \ldots, r \tag{4.7}$$

where $\bar{L}_j$ is the feedback matrix of the jth fuzzy rule. This control law is designed like a LQR (Linear Quadratic Regulator) [19] to minimize a performance index:

$$J_j(x, u) = \sum_{k=0}^{\infty} [x^T(k) \bar{Q}_j x(k) + u^T(k) R_j u(k)] \qquad j = 1, \ldots, r \tag{4.8}$$

where $\bar{Q}_j = \bar{Q}_j^T \geq 0$ and $R_j = R_j^T \geq 0$. The control design by LQR for each local model requires the algebraic solution of the Ricatti equation for the H_j matrix.

$$A_j^T H_j A_j - H_j + \bar{Q}_j - A_j^T H_j B_j (R_j + B_j^T H_j B_j)^{-1} B_j^T H_j A_j = 0 \qquad (4.9)$$

So, the feedback matrices are calculated like:

$$\bar{L}_j = (R_j + B_j^T H_j B_j)^{-1} B_j^T H_j A_j \quad j = 1, \ldots, r \qquad (4.10)$$

The closed loop system is:

$$\begin{aligned} x(k+1) &= \sum_{i=1}^{r} \sum_{j=1}^{r} \alpha_i \beta_j (A_i - B_i \bar{L}_j) x(k) \qquad (4.11) \\ &= \sum_{i=1}^{r} \sum_{j=1}^{r} \alpha_i \beta_j \Lambda_{ij} x(k) \end{aligned}$$

with $\Lambda_{ij} = A_i - B_i \bar{L}_j$, $i = 1, \ldots, r$ and $j = 1, \ldots, r$.

The following properties of the antecedent part (Eq. 2.7) are considered for the stability analysis of fuzzy control (Eq. 4.7):

$$h_i h_j \geq 0 \quad \sum_{i=1}^{r} \sum_{j=1}^{r} h_i h_j = 1 \quad \sum_{i=1}^{r} h_i^2 + 2 \sum_{i,j}^{i<j} h_i h_j = 1 \qquad (4.12)$$

Based on the properties of fuzzy control, and assuming two-overlapped fuzzy memberships at most, the stability analysis of closed loop fuzzy control is presented.

Theorem 4.1 *There is a controller in discrete time such that $x_c \to 0$ and $x_p \to 0$ when $k \to \infty$, if there are two matrices defined positive $P_1 \in \Re^{n \times n}$ and $P_2 \in \Re^{n \times n}$, such that the following conditions of the LMIs hold:*

$$(A_j^p - B_j^p L_j^p)^T P_1 (A_j^p - B_j^p L_j^p) - P_1 < 0 \qquad (4.13)$$

$$(A_j^c - B_j^c L_j^c)^T P_2 (A_j^c - B_j^c L_j^c) - P_2 < 0 \qquad (4.14)$$

In the case of a fault event disruption, the error (as well as time delays) modifies the local structure through time, as an overall situation. Although ρ_j^p or ρ_j^c in Eq. 4.3 highlight a missing state in terms either plant or controller as described in Chap. 2. The proposal of a current group of observers is to guarantee system structure. Therefore, stability, in terms of time delays and misleading structure, should be accomplished. The related observers states are presented as $z(k)$ following [20]. In this case N_j, M_j, L_j, and T_2 are the observer parameters to be defined as a classic Unknown Input Observer (UIO).

The UIO is represented as:

$$z(k+1) = \sum_{j=1}^{r} h_j(N_j z(k) + M_j u_1^p(k - \hat{\tau}_{1j}^{sc}) + L_j y_p(k)) \tag{4.15}$$

$$\hat{x}_p(k) = z(k) + T_2 y_p(k) \tag{4.16}$$

The objective now is to find the observer's gains E, G, N_j, M_j, L_j, and T_2, this is, to design the observer such that the observer error $e_o(k) \to 0$ when $k \to \infty$. Thus, defining $e_o(k) = x_p(k) - \hat{x}_p(k)$, and substituting Eq. 4.16:

$$e_o(k) = x_p(k) - z(k) - T_2 C x_p(k) - T_2 G u_2^p(k - \hat{\tau}_{2j}^{sc}) \tag{4.17}$$

Assuming there is a T_1 and T_2 with the following constraints:

$$T_1 E + T_2 \sum_{i=1}^{r} C_i = I_n \tag{4.18}$$

With the constraints (4.18), the observer error is:

$$e_o(k) = T_1 x_p(k) - z(k) \tag{4.19}$$

and its discrete difference is:

$$e_o(k+1) = T_1 x_p(k+1) - z(k+1) \tag{4.20}$$

Substituting $x_p(k+1)$ and $z(k+1)$ using Eqs. 2.12 and 4.15, respectively:

$$\begin{aligned} e_o(k+1) = \sum_{j=1}^{r} h_j[T_1(A_j^p x_p(k) + \rho_j^p B_{1j}^p u_1^p(k - \hat{\tau}_{1j}^{sc}) + B_{2j}^p u_2^p(k - \hat{\tau}_{2j}^{sc})) \\ - N_j z(k) - M_j u_1^p(k - \hat{\tau}_{1j}^{sc}) - L_j y_p(k)] \end{aligned} \tag{4.21}$$

Reordering, and using the observer error definition:

$$\begin{aligned} e_o(k+1) = \sum_{j=1}^{r} h_j[N_j e_o(k) + \left(T_1 A_j^p - N_j T_1 - L_j \sum_{i=1}^{r} C_i\right) x_p(k) \\ + (T_1 B_{1j}^p - M_j) u_1^p(k - \hat{\tau}_{1j}^{sc}) + T_1 B_{2j}^p u_2^p(k - \hat{\tau}_{2j}^{sc})] \end{aligned} \tag{4.22}$$

In terms of augmented states, the estimated states are presented by the observer, since the local variable structure is used when local fault appears as well as the time delays.

Let the augmented state be the union of the fuzzy model, the fuzzy controller, and the fuzzy observer:

$$\begin{bmatrix} x_p(k+1) \\ e_o(k+1) \\ \hat{x}_p(k+1) \end{bmatrix} = \sum\nolimits_{j=1}^{r} h_j$$
$$\begin{bmatrix} A_j^p x_p(k) + \rho_j^p B_{1j}^p u_1^p(k) + \rho_j^p B_{2j}^p u_2^p(k) \\ N_j e_o(k) + (T_1 A_j^p - N_j T_1 - L_j \sum_{i=1}^{r} C_i) x_p(k) + (T_1 B_{1j}^p - M_j) u_1^p(k) + T_1 B_{2j}^p u_2^p(k) \\ A_j^c x_c(k) + \rho_j^c B_{1j}^c u_1^c(k) + \rho_j^c B_{2j}^c u_2^c(k) \end{bmatrix} \tag{4.23}$$

Reordering this last equation, and using a matrix representation:

$$\begin{aligned} \begin{bmatrix} x_p(k+1) \\ x_c(k+1) \\ e_o(k+1) \end{bmatrix} &= \sum_{j=1}^{r} h_j \begin{bmatrix} A_j^p & 0 & 0 \\ 0 & A_j^c & 0 \\ T_1 A_j^p + N_j T_1 - L_j \sum_{i=1}^{r} C_i & 0 & N_j \end{bmatrix} \begin{bmatrix} x_p(k) \\ x_c(k) \\ e_o(k) \end{bmatrix} \\ &+ \sum_{j=1}^{r} h_j \begin{bmatrix} \rho_j^p B_j^p & 0 \\ 0 & \rho_j^c B_j^c \\ T_1 B_{1j}^p - M_j & 0 \end{bmatrix} \begin{bmatrix} u_1^p(k - \hat{\tau}_{1j}^{sc}) \\ u_1^c(k - \hat{\tau}_{1j}^{ca}) \end{bmatrix} \\ &+ \sum_{j=1}^{r} h_j \begin{bmatrix} 0 & 0 \\ 0 & 0 \\ T_1 B_{2j}^p & 0 \end{bmatrix} \begin{bmatrix} u_2^p(k - \hat{\tau}_{2j}^{sc}) \\ u_2^c(k - \hat{\tau}_{2j}^{ca}) \end{bmatrix} \end{aligned} \tag{4.24}$$

The observer error of the system is:

$$\begin{bmatrix} x_p(k+1) \\ x_c(k+1) \\ e_o(k+1) \end{bmatrix} = \sum_{j=1}^{r} h_j \begin{bmatrix} \Lambda_j^p & 0 & 0 \\ 0 & \Lambda_j^c & 0 \\ 0 & 0 & N_j \end{bmatrix} \begin{bmatrix} x_p(k) \\ x_c(k) \\ e_o(k) \end{bmatrix} \tag{4.25}$$

where $\Lambda_j^p = A_j^p - B_{1j}^p L_j^p$ *and* $\Lambda_j^c = A_j^c - B_{1j}^c L_j^c$.

Defining a parameter $Q_j = N_j T_2 - L_j$*, and substituting the feedback Eq. 4.7 for plant and controller:*

$$\begin{bmatrix} x_p(k+1) \\ x_c(k+1) \\ e_o(k+1) \end{bmatrix} = \sum_{j=1}^{r} h_j \begin{bmatrix} A_j^p - B_{1j}^p L_j^p & 0 & 0 \\ 0 & A_j^c - B_{1j}^c L_j^c & 0 \\ 0 & 0 & T_1 A_j + Q_j C_j \end{bmatrix} \begin{bmatrix} x_p(k) \\ x_c(k) \\ e_o(k) \end{bmatrix} \tag{4.26}$$

This is so if the following constraints are fulfilled:

$$I_n = T_1 + T_2 \sum_{i=1}^{r} C_i \tag{4.27}$$

$$0 = T_1 B_{2j} \tag{4.28}$$

$$M_j = T_1 B_{1j} \tag{4.29}$$

$$L_j = N_j T_2 - Q_j \tag{4.30}$$

$$N_j = T_1 A_j + Q_j \sum_{i=1}^{r} C_i \tag{4.31}$$

Verifying the constraints of Eqs. 4.27, 4.28, and 4.31, the problem reduces to find a matrix Θ such that:

$$\Theta X = Y \tag{4.32}$$

$$N_i = \Theta Y_i \tag{4.33}$$

where Θ, X, Y, and Y_i are defined as follows:

$$\Theta = [T_1 T_2 Q_1 Q_2 \ldots Q_r] \tag{4.34}$$

$$X = \begin{bmatrix} 1 & B_{21}^p & \ldots & B_{2r}^p \\ \sum_{i=1}^{r} C_i & 0_{m\times q} & \ldots & 0_{m\times q} \end{bmatrix} \tag{4.35}$$

$$Y = [I_n 0_{n\times rq}] \tag{4.36}$$

$$Y_i = \begin{bmatrix} A_i \\ e_i \otimes \sum_{i=1}^{r} \end{bmatrix} \tag{4.37}$$

where e_i is the column vector with all its components equal to zero except the i-th component, which is 1.

Lemma 4.1 *There is a UIO (Eqs. 4.15 and 4.16) such that the following conditions stands:*

$$rank\ X = rank\ \begin{bmatrix} B_{21}^p & \ldots & B_{2r}^p \end{bmatrix} + n \tag{4.38}$$

Once defined the existence of the UIO, it is necessary to define the observability of the system. The following lemma states the observability conditions.

Lemma 4.2 *The estimation error of the UIO tends to zero ($e_o \rightarrow 0$) while time tends to infinity ($k \rightarrow \infty$) if there is a matrix $P \in \Re^{n\times n}$, and a matrix $\hat{Z} \in \Re^{n\times(n+m(r+1))}$ such that the following LMIs are fulfilled for $i = 1 \ldots r$:*

$$\begin{bmatrix} \Phi_i & (\bar{X}Y_i)^T \hat{Z}^T \\ \hat{Z}(\bar{X}Y_i) & -P \end{bmatrix} < 0 \tag{4.39}$$

where matrix $\bar{X} = 1 - XX^+$, and X^+ is the pseudo inverse of matrix X, and Φ_i is:

$$\Phi_i = (\bar{X}Y_i)^T \hat{Z}^T (X^+ Y_i) + (YX^+Y_i)^T Z(\bar{X}Y_i) + (YX^+Y_i)^T P(YX^+Y_i) - P \tag{4.40}$$

The complete conditions of the controllers and the UIO are:

$$(A_j^p - B_{1j}^p L_j^p)^T P_1 (A_j^p - B_{1j}^p L_j^p) - P_1 < 0 \quad (4.41)$$

$$(A_j^c - B_{1j}^c L_j^c)^T P_2 (A_j^c - B_{1j}^c L_j^c) - P_2 < 0 \quad (4.42)$$

$$N_i^T P_3 N_i - P_3 < 0 \quad (4.43)$$

These provide the base for the following theorems.

Theorem 4.2 *There is a UIO in discrete time, if the condition in Eq. 4.38 holds, and if there is a matrix $P_3 \in \Re^{n \times n}$, and a matrix $\hat{Z} \in \Re^{n \times (n+m(r+1))}$ such that there exist the LMIs in Eq. 4.39.*

For designing the controllers and the UIO in discrete time, the next procedure is followed:

1. *Design the controller for the plant L_j^p such that the Theorem 4.1 holds.*
2. *Verify the condition of Lemma 4.1.*
3. *Solve the LMIs for P_3 and $\hat{Z}$ (Eq. 4.39).*
4. *Obtain Z with $Z = P^{-1}\hat{Z}$.*
5. *Obtain N_j, M_j, and L_j (Eqs. 4.29, 4.30 and 4.31).*

Now, the related information amongst local plants and controllers has been reviewed up to this point. Additionally, it is important to consider the nature of time delays, along with a review of the local observers. Here, the time delays are independent, based on the time obtained from the scheduling approximation:

$$\tau_i^{sc} + \tau_i^{ca} < T \qquad i = 1, 2, \ldots \quad (4.44)$$

Further, solving this condition is numerically possible through an LMI solving strategy like [7]. However, the current system representation results quite restrictive, although possible, by using Fuzzy TSK where variables from fuzzy structure control design and observers definitions are tuned through this strategy.

4.3 Sampling Frequency Control

An approximation to schedule a real-time distributed system is based on bounding time delays through modifications on frequency transmission of individual components in the system, as presented in [14]. From this, the scheduling of a distributed system can be accomplished through modifications on transmission frequencies into a region, where the system performance is not affected. This approach drives the periodicity transmission though three parameters: minimum frequency f_m, real frequency f_r, and maximum frequency f_x. The distributed system dynamics can be modeled as a linear time-invariant system, whose state variables are transmission frequencies $f_i = \frac{1}{p_i}$ of the n nodes involved in the NCS. Notice that there is a relationship between node frequencies and external input frequencies, which, therefore, serves as coefficients for the linear system. The objective of controlling periodicity is to achieve coordination through the convergence of values. Each sensor i has a minimum f_m^i and maximum frequencies f_x^i transmission rates, which are computed off-line. Using the model presented in Sect. 2.3, a control law can be designed to modify the sampling frequencies.

The control input is given by a function of desired frequencies (f_d) and real frequencies (f_r) for all the sensors:

$$u = -k^s(f_d - f_r) = -k^s x(k) \tag{4.45}$$

Note that k^s is the control gain, defined as the basics of a LQR algorithm, and f_d and f_r are vectors. Now, the plant 2.14 is defined as follows, in terms of frequency transmission:

$$\begin{aligned} x(k+1) &= Ax(k) + Bu(k) \\ x(k+1) &= A(f_d - f_r)(k) - Bk^s(f_d(k) - f_r(k)) \\ x(k+1) &= Af_d(k) - Af_r(k) - Bk^s f_d(k) + Bk^s f_r(k) \\ x(k+1) &= (A - Bk^s)(f_d - f_r)(k) \end{aligned} \tag{4.46}$$

In this case, the network constraint (2.13) needs to be accomplished. Regarding the network utilization, this equation implies that consumptions due to message transmission should not exceed the transmission periods, on order to avoid packet losses and time delays.

Defining $e(k) = (f_d - f_r)(k)$, Eq. 4.46 becomes:

$$e(k+1) = (A - Bk^s)e(k) \tag{4.47}$$

Now defining $e_5(k) = \varepsilon_d(k) - \varepsilon_r(k)$ as the augmented state, the system is as follows:

$$\begin{bmatrix} e_1(k+1) \\ e_2(k+1) \\ e_3(k+1) \\ e_4(k+1) \\ e_5(k+1) \end{bmatrix} = \begin{bmatrix} \frac{\bar{\lambda}}{f_m^1} & \frac{f_m^2}{f_m^1} & \frac{f_m^3}{f_m^1} & \frac{f_m^4}{f_m^1} & 0 \\ \frac{f_m^1}{f_m^2} & \frac{\bar{\lambda}}{f_m^2} & \frac{f_m^3}{f_m^2} & \frac{f_m^4}{f_m^2} & 0 \\ \frac{f_m^1}{f_m^3} & \frac{f_m^2}{f_m^3} & \frac{\bar{\lambda}}{f_m^3} & \frac{f_m^4}{f_m^3} & 0 \\ \frac{f_m^1}{f_m^4} & \frac{f_m^2}{f_m^4} & \frac{f_m^3}{f_m^4} & \frac{\bar{\lambda}}{f_m^4} & 0 \\ c_1 & c_2 & c_3 & c_4 & 1 \end{bmatrix} -$$

$$\begin{bmatrix} f_x^1 & 0 & 0 & 0 & 0 \\ 0 & f_x^2 & 0 & 0 & 0 \\ 0 & 0 & f_x^3 & 0 & 0 \\ 0 & 0 & 0 & f_x^4 & 0 \\ 0 & 0 & 0 & 0 & 1 \end{bmatrix} \begin{bmatrix} k_1^s & 0 & 0 & 0 & 0 \\ 0 & k_2^s & 0 & 0 & 0 \\ 0 & 0 & k_3^s & 0 & 0 \\ 0 & 0 & 0 & k_4^s & 0 \\ 0 & 0 & 0 & 0 & k_5^s \end{bmatrix} \begin{bmatrix} e_1(k) \\ e_2(k) \\ e_3(k) \\ e_4(k) \\ e_5(k) \end{bmatrix} \tag{4.48}$$

$$\begin{bmatrix} e_1(k+1) \\ e_2(k+1) \\ e_3(k+1) \\ e_4(k+1) \\ e_5(k+1) \end{bmatrix} = \begin{bmatrix} \frac{\bar{\lambda}}{f_m^1} - f_x^1 k_1^s & \frac{f_m^2}{f_m^1} & \frac{f_m^3}{f_m^1} & \frac{f_m^4}{f_m^1} & 0 \\ \frac{f_m^1}{f_m^2} & \frac{\bar{\lambda}}{f_m^2} - f_x^2 k_2^s & \frac{f_m^3}{f_m^2} & \frac{f_m^4}{f_m^2} & 0 \\ \frac{f_m^1}{f_m^3} & \frac{f_m^2}{f_m^3} & \frac{\bar{\lambda}}{f_m^3} - f_x^3 k_3^s & \frac{f_m^4}{f_m^3} & 0 \\ \frac{f_m^1}{f_m^4} & \frac{f_m^2}{f_m^4} & \frac{f_m^3}{f_m^4} & \frac{\bar{\lambda}}{f_m^4} - f_x^4 k_4^s & 0 \\ c_1 & c_2 & c_3 & c_4 & 1 - k_5^s \end{bmatrix} \begin{bmatrix} e_1(k) \\ e_2(k) \\ e_3(k) \\ e_4(k) \\ e_5(k) \end{bmatrix} \tag{4.49}$$

Hence, the local application of the scheduling algorithm is needed for the partial recovery of the NCS, but this implies that the overall performance of the NCS is now degraded.

4.4 Control—Scheduling Codesign

Codesign has the main objective of mixing the advantages of two or more knowledge areas, while minimising their disadvantages, aiming to adequately solve a common problem. In NCS, the control-communication codesign commonly integrates the feedback control and real-time communications. It is based on the principle that the performance of the system depends on the design of the control algorithms, as well as on the performance of the scheduling of shared communication resources.

Unfortunately, NCS design is frequently based on the principle of separation of concerns [21]. This principle assumes that feedback controllers can be modelled and implemented as periodic tasks with a fixed period, and with a known consumption, for instance, Worst Case Execution Time (WCET), and a rigid deadline. These assumptions focus on the control problem without worrying about how tasks are scheduled in the nodes, and how packets are sent through the network. Nevertheless, it is not quite understood how scheduling impacts on the system's performance.

The control of tasks does not always make use of the available computing resources in an optimal way, and the assumptions of the model of a simple task are too restrictive regarding the characteristics of many control systems. For example, many deadlines are not always rigid, this is, many practical control systems can occasionally tolerate losing their deadlines. To solve the limitation of resources in a NCS, codesign is made necessary at several different levels. For instance, hardware/software codesign, mechanic/electric codesign, etc. Here, the control/scheduling codesign is presented. In practical terms, the control/communication codesign can be divided into two categories: feedback control of computational systems and real-time control.

The control of computational systems is also known as feedback scheduling [21]. The basic idea is dealing with the scheduling problem as a feedback control problem. A control feedback loop is introduced in the resource scheduler of the computational systems. The objective of feedback scheduling is to increase the flexibility with respect to the uncertainties in resource utilisation. Instead of pre-assigning resources based on an off-line analysis, resources are assigned dynamically on-line, based on a feedback from the actual utilisation of the resource.

A second category of codesign focus on real-time control. Branisky et al. [22] is perhaps one of the first authors to establish that scheduling algorithms and control systems cannot be separately designed. They work the problem of plant regulation as an optimal scheduling problem, studying the limitations of Rate Monotonic (RM) scheduling, as well as the limitations on NCS stability, taking into consideration issues such as induced time delays, packet loss, and multiple packet transmission. Branisky [22] considers a set of NCSs as linear plants with transmission only between sensor and controller, with a period T equal to its deadline, and with a consumption time c_i. Applying the RM algorithm for scheduling, a static priority is assigned to each controlled plant. A higher priority is assigned to a plant with faster dynamics since it has a higher transmission rate than other slower dynamic plants. The set of N tasks is feasible if the utilisation factor of the network U is less than 1. For optimising the scheduling, it is assumed that each NCS is associated with an average performance function h, which provides with the cost of control as a function of the transmission period h_i. Selecting such a performance function is essential for the optimisation problem. Normally, a quadratic or exponential cost is chosen. Branisky et al. [22] analyse the allowed percentage of packet loss that secures the stable performance of a NCS. The NCS with packet loss is modelled as a dynamic asynchronous system. It is assumed that the sampling period is constant, and the NCS tolerates a certain amount of lost feedback data. Even though they employ the concept of codesign of a NCS, the method has a few liabilities due to they only consider the data transmission between sensor and controller, and a constant time delay is considered for the analysis of lost packets.

In [23] it is made use of the concept of feedback control of computational resources for scheduling the resources and the workload of a processor. They propose two schemes to manage the systems with a control task or multitask. The design of the control system takes into account unknown time delays, given some uncertainties of time, which are unavoidable. Likewise, they present a new control design method, using state feedback for discrete systems with time delays, formulated as LMI. For feedback scheduling, they consider communication delays mainly as the latency input-output. The objective is to on-line adjust the sampling period of the controller, to cope with the requirements of computational resources, using discrete time systems, in which, for calculating the control input, the communication delay is obtained by the addition of the network induced delay plus the computational cost delay.

Fuzzy Controller

A new modelling method for nonlinear NCS with variable time delays is presented in Sect. 2.4. It is represented by a TSK fuzzy model [15], with the estimated time delay $\hat{\tau}(k)$ as input of the antecedent part, and linear discrete models with different sampling periods h_j as consequent part. Once designed the fuzzy model a fuzzy controller is proposed. Thus, defining r fuzzy rules, the jth rule has the form:

$$if\ \hat{\tau}(k)\ is\ \alpha_j(\hat{\tau})\ then\ u_j = \Phi(h_j)K_j\hat{x}_k \tag{4.50}$$

where $x(k) \in \Re^n$ is the state vector of system, $u(k) \in \Re^n$ is the input vector of process, $\alpha_j(\hat{\tau})$ is the jth membership function of the estimated time delay $\hat{\tau}(k)$.

The overall fuzzy feedback control is:

$$u_k = -\sum_{j=1}^{r} \psi_j K_j \hat{x}_k \quad j = 1, \ldots, r \tag{4.51}$$

where K_j is the feedback matrix of the *j-th* fuzzy rule. This control law is designed like a LQR (Linear Quadratic Regulator) [19] to minimize a performance index. The control design by LQR for each local model requires the algebraic solution of the Ricatti equation for the H_j matrix. So, the feedback matrices are calculated like:

$$K_j = \left(R_j + \Gamma_j^T H_j \Gamma_j\right)^{-1} \Gamma_j^T H_j \Phi_j \quad j = 1 - r \tag{4.52}$$

The closed loop system is:

$$\begin{aligned} x_{k+1} &= \sum_{i=1}^{r} \sum_{j=1}^{r} \alpha_i \beta_j \left(\Phi_i - \Gamma_i K_j\right) x_k \\ &= \sum_{i=1}^{r} \sum_{j=1}^{r} \alpha_i \beta_j \Lambda_{ij} x_k \end{aligned} \tag{4.53}$$

with $\Lambda_{ij} = \Phi_i - \Gamma_i K_j \quad i = 1, \ldots, r \quad j = 1, \ldots, r$.

The following properties of the antecedent part are considered for the stability analysis of fuzzy control (4.51):

$$\begin{aligned} &\psi_i \psi_j \geq 0 \\ &\sum_{i=1}^{r} \sum_{j=1}^{r} \psi_i \psi_j = 1 \\ &\sum_{i=1}^{r} \psi_i^2 + 2 \sum_{i,j}^{i<j} \psi_i \psi_j = 1 \end{aligned} \tag{4.54}$$

Using this properties and assume that two-overlapped fuzzy memberships at most a stability analysis is presented in [24] using the Theorem 4.1.

Theorem 4.3 *The equilibrium state $x_e = 0$ of closed loop system with control input (4.53 with two-overlapped fuzzy memberships at most, is asymptotically stable in the large, if there exist μ positive definite matrices $P_s = P_s^T > 0$ such that:*

$$\begin{gathered} \left(\Lambda_{ij}^s + \Lambda_{ji}^s\right)^T P_s \left(\Lambda_{ij}^s + \Lambda_{ji}^s\right) - 2P_s < 0 \\ i \in S_s \\ s = 1, \ldots, \mu \end{gathered} \tag{4.55}$$

$$\begin{aligned} \left(\Lambda_{ii}^{s}\right)^{T} P_s \Lambda_{ii}^{s} - P_s < 0 \\ i \in S_s \\ j \in S_s \\ i < j \in S_s \end{aligned} \tag{4.56}$$

$\Lambda_{ij} = \Phi^i - \Gamma^i K^j$ where $S = \{S_1, S_2, \ldots, S_\mu\}$ are μ regions where two fuzzy rules are fired (overlapped fuzzy memberships) at most, where S_s contains the membership function indexes for fired fuzzy rules in s *region.*

Scheduling

In this section, scheduling theory is used for designing a feedback scheduler that, online, assigns an execution period for the NCS, based on the control system's performance and the load conditions of the network. The change in the sampling period is carried out depending on a scheduling period. The basic idea is to keep the lost deadlines (those packets that overpass the deadline or are simply lost), defined as the Quality of Service (QsS), and the system's performance (QsC) on the desired level, adjusting the sampling period. Here, it is proposed a local fuzzy scheduler for each sensor node present in the communication network, based on external traffic performs a dynamic scheduling, also known as feedback scheduling. This feedback scheduler is codesigning with the Fuzzy controller in the Sect. 4.4.

The configuration of the proposed dynamic scheduling is shown in Fig. 4.4, whose behaviour is described as follows:

1. The sensor sends a packet with period h to the controller. The packet contains system's information along with its execution period.
2. The controller adds the error and control signals as part of the packet and sends it to the actuator.
3. The actuator calculates QsC and QsS for each scheduling period and sends the packet to the sensor.
4. Finally, the sensor modifies the period h based on QsC and QsS. A lost deadline occurs when a packet arrives at the actuator *after* the deadline, or when a packet is lost.

In control terms, the dependent variable is h, whereas the controlled variables are QsS and QsC.

The actuator computes the system's performance Δe using the Mean Absolute Error (MAE) of n packet, received within a scheduling period δ:

$$\Delta e = \frac{\sum_{k \in n} |e_k|}{n} \tag{4.57}$$

On the other hand, the lost packets rate Δh is obtained using the information of the total of received packets m that have exceeded their deadline h_{max} plus the lost packets λ during the scheduling period:

$$\Delta h = \frac{m}{\delta} \sum_{k \in n} h_k \qquad m = \{\forall k \in n | \tau_k > h_{max}\} + \lambda \tag{4.58}$$

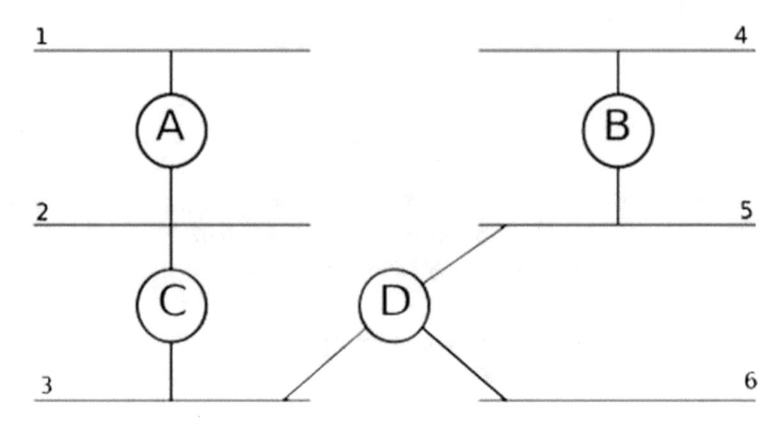

Fig. 4.4 Codesign configuration

where λ is the number of lost packets during the scheduling period, τ_k is the time delay between sensor and actuator, measured during the scheduling period, and h_{max} is the deadline for the received packet.

The election of the lost deadlines rate, as the controlled variable, is due to it is one of the most common metrics used for Quality of Service (QoS). From a real-time point of view, it is an important factor that affects QoC. The QoS keeps controlled the lost deadline rate in an acceptable low level. Besides, using Δh as the controlled variable allows to simultaneously address variable time delays, as well as packet loss.

When Δh is kept at a low level, the time delay of most packets is less than the deadline, and the number of lost packets is bounded. As a consequence, the impact on QsC of the time delay and the lost packets diminishes.

The sampling period affects the lost deadlines rate: with a short sampling period, the used of the network increases, which inherently causes an increase of network uncertainties, and vice versa; with a heavy load of traffic, the collision probability between nodes increases, and potentially, grows the time delays and the packet loss, which at the same time increase the lost deadlines. Usually, a largely lost deadline rate may be reduced by increasing the sampling period, particularly when the network is overloaded.

From Digital Control Theory for sampled data, short sampling periods improves QoC. Here, QoC increases by increasing the efficient use of the network, by adjusting the sampling period. This justifies the election of the sampling period as the dependent variable, which is adjusted depending on network conditions, where variations of unpredictable and dynamic traffic in the NCS may be effectively compensated.

With metrics for QsC (Δe) and QsS (Δh), a fuzzy scheduler is designed as a control problem in which both metrics compose the IF rules, while the THEN rules are parallel feedback laws. The i-th fuzzy rule has the following form:

$$Rule\ i : IF\ \Delta e\ is\ \beta_1\ and\ \Delta h\ is\ \beta_2\ THEN\ \overline{h} = Fz \tag{4.59}$$

where β_i are the membership functions, and $z \in \Re^2$ with $z = [\Delta e \;\; \Delta h]$, and F is the scheduling feedback matrix. The complete fuzzy scheduler is then:

$$\overline{h} = \sum_{i=1}^{M} \prod_{j=1}^{2} \beta_{ij} F_i z \tag{4.60}$$

where M is the number of fuzzy rules. The new sampling period is assigned to the interval:

$$h^{+} = \{h_{min} \leq \overline{h} \leq h_{max}\} \tag{4.61}$$

Given the absence of a mathematical model describing the relationship between lost deadline rate, the system's MAE, and the sampling period, the scheduling parameters are set based on simulation.

Any strategy to design a control law needs to take into account a gain scheduling approximation. To do so, a fuzzy control strategy based on Takagi-Sugeno-Kang (TSK) is applied. Using fuzzy control systems [16], and considering a variable structure system representation.

Based on the developments in Chaps. 2 and 4, where the time delays have been extensively worked, it seems important that this representation is useful for conditioning the time delay in terms of a mobile environment, such as the one produced by consensus or context changes in communications. Thus, the time delay representation in terms of a period function or as a product of a stochastic scenario allows concluding in a similar way to the Theorems 4.1 and 4.3. Hence, the time delay is still locally and globally bounded.

The authors consider that even in extreme cases, in which communications are lost, the same theory developed in Chaps. 2 and 4 sustain through packet loss. A scenario of possible interest for the reader is the extensive use of the communication channel, given the local need to establish a group context at each node. This increases the traffic, which conveys to a known situation and modelled before.

4.5 Concluding Remarks

In this chapter we provide a review of the classical control methods for NCS, presenting strategies implemented successfully in real systems. In addition, three controllers are presented using the NCS models of Chap. 2.

The fuzzy control methodology allows generating an optimal control signal in function of the imperfections of the network through its prediction. Generating an optimal control signal for the magnitude of the imperfection while ensuring the global stability of the system.

The sampling frequency scheduling allows minimising network usage, increasing the resources for nodes that need to improve their transmission in the presence of disturbances. The result is a scheduler that modifies the frequencies of each node in the range of its minimum and maximum frequencies, adapting to the network and control performances.

Finally, the codesign strategy employs two fuzzy models. One model with variable actuation periods as input to generate an optimal control signal in function of the next actuation period and another model to schedule the next actuation period in function of QsC and QsS to simultaneously control the system in function of the imperfections of the network.

References

1. Kim, K.D., Kumar, P.R.: Real-time middleware for networked control systems and application to an unstable system. Control Syst. Technol. IEEE Trans. **21**(5), 1898–1906 (2013)
2. Fridman, E., Niculescu, S.I.: On complete Lyapunov-Krasovskii functional techniques for uncertain systems with fast-varying delays. Int. J. Robust Nonlinear Control **18**(3), 364–374 (2008)
3. Gupta, R.A., Chow, M.Y.: Networked control system: overview and research trends. Ind. Electron. IEEE Trans. **57**(7), 2527–2535 (2010)
4. Nilsson, J.: Real-Time Control Systems with Delays; PhD Thesis, Dept. Automatic Control. Lund Institute of Technology, Lund, Sweden (1998)
5. Benitez-Perez, H., Benitez-Perez, A., Ortega-Arjona, J.: Networked control systems design considering scheduling restrictions and local faults using local state estimation. IJICIC **9–8**, 3225–3239 (2013)
6. Tanaka, K., Sugeno, M.: Stability analysis and design of fuzzy control systems. Fuzzy Sets Syst. **45**(2), 135–156 (1992)
7. Tanaka, K., Ikeda, T., Wang, H.: Design of fuzzy control systems based on relaxed LMI stability conditions. In: Proceedings of the 35th IEEE Conference on Decision and Control, pp. 598–603. IEEE (1996)
8. Goktas, F., Smith, J.M., Bajcsy, R.: μ-Synthesis for distributed control systems with network-induced delays. In: Proceedings of the 35th IEEE Conference on Decision and Control **1**, pp. 813–814. IEEE (1996)
9. Takaba, K.: Robust preview tracking control for polytopic uncertain systems. In: IEEE Conference on Decision and Control **2**, pp. 1765–1770. IEEE (1998)
10. Olshusky, A., Tsitsiklis, J.: Degree fluctutions and the convergence time of consensus algorithms. IEEE Trans. Autom. Control **58–10**, 2626–2631 (2013)
11. Walsh, G.C., Ye, H., Bushnell, L.G.: Stability analysis of networked control systems. Control Syst. Technol. IEEE Trans. **10**(3), 438–446 (2002)
12. Mendez-Monroy, E.: Codiseño de Sistemas de Control en Red Compensando Imperfecciones Acotadas de Tiempo Inducidos por la Red. Posgrado en Ingenieria, Campo de Conocimiento de Electrica, PhD, UNAM. 22 Junio (2012)
13. Halevy, Y., Ray, A.: Integrated communication and control systems: Part I analysis. J. Dyn. Syst. Meas. Control, **110** 367–373 (1988)
14. Esquivel-Flores, O., Benitez-Perez, H.: Reconfiguracion Dinamica de Sistemas Distribuidos en Tiempo Real basada en Agentes. Revista Iberoamericana de Automatica e Informatica Industrial **9–3**, 300–313 (2012)
15. Tanaka, K., Wang, H.O.: Fuzzy Control Systems Design and Analysis: A Linear Matrix Inequality Approach. Wiley (2001)
16. Mendez-Monroy, P.E., Benitez-Perez, H.: Supervisory fuzzy control for networked control systems. Int. J. Innov. Comput. Inf. Control Express Lett. **3–2**, 233–240 (2009)

17. Liu, C., Layland, J.: Scheduling algorithms for multiprogramming in a hard-real-time environment. J. Assoc. Comput. Mach. **20–1**, 46–61 (1973)
18. Buttazzo, G.C.: Real Time Systems. Springer (2004)
19. Zhang, H., Yang, D., Chai, T.: Guaranteed cost networked control for T-S fuzzy systems with time delays. IEEE Trans. Syst. Man Cybern. C **37–2**, 160–172 (2007)
20. Zhu, X., Hua, C., Wang, S.: State feedback controller design of networked control systems with time delay in the plant. Int. J. Innov. Comput. Inf. Control **4**(2), 283–290 (2008)
21. Arzen, K.E., Bernhardsson B., Eker J., Cervin A, Nilsson K., Persson P., Sha, L.: Integrated control and scheduling. Technical Report ISRN LUTFD2/TFRT7586SE, Lund Institute of Technology, Sweden (1999)
22. Branisky, M.S., Phillips, S.M., Zhang, W.: Scheduling and feedback co-design for networked control systems. Proc. IEEE Conf. Decis. Control **2**, 1211–1217 (2002)
23. Sename, O., Simon, D., Robert, D.: Feedback scheduling for real-time control of systems with communication delays. In: Proceedings of the IEEE Conference on Emerging Technologies and Factory Automation (2003)
24. Mendez-Monroy, P.E., Benitez-Perez, H.: Fuzzy control with time delay estimation for networked control systems. In: 2009 6th International Conference on Electrical Engineering, Computing Science and Automatic Control (CCE) (2009). https://doi.org/10.1109/ICEEE.2009.5393457

Chapter 5
Control Design Considering Mobile Computing

Abstract In this chapter, an extension of distributed systems considering mobile computing is reviewed. First, A review of some scheduling algorithms is presented to get a good approximation to bounded time delays considering the spend time in all stages of communication and compute. A review of the most important algorithms in terms of consensus and routing is presented. A computer network design is presented from a selection of real-time features, scheduler, task handlers, priorities, precedencies and consensus. Finally, a brief review of control design for mobile conditions is presented with the most representative algorithms in the literature.

5.1 Dynamic Behaviour onto Real Time

Real-time strategies convert the idea of bounded response systems into a dynamic operation, considering the certainty onto time response. To achieve such a goal, the use of scheduling algorithms is preferable. A real-time system is where the interaction amongst elements are with a certainty of time response as global and local through the processing elements. The most common algorithms for real-time are the schedulers, like Earliest Deadline First or Priority Exchange, that manage those tasks in preemptive mode [1].

In particular, one of the most important properties of hard real-time systems is predictability [2]. This means that, based on features of the kernel and the information associated with each task, the system is able to predict the evolution of its tasks and to successively guarantee that all time restrictions are accomplished. However, this is possible only occasionally, due to the high degree of non-determinism of the distributed systems. Several factors introduce scheduling uncertainties, such as the features of the processor and the kernel, the applied scheduling algorithms, the synchronisation mechanisms, the memory management, the communications, the interruption management mechanisms, among others [3].

Distributed systems whose functionality is restricted to accomplishing time deadlines are known as Real-time distributed systems (RTDS), as reviewed in the previous section. These constitute the object study of the present book. Due to the increasing complexity of the interaction of the elements of a RTDS, it is necessary to carry out

H. Benítez-Pérez et al., *Control Strategies and Co-Design of Networked Control Systems*, Modeling and Optimization in Science and Technologies 13,
https://doi.org/10.1007/978-3-319-97044-8_5

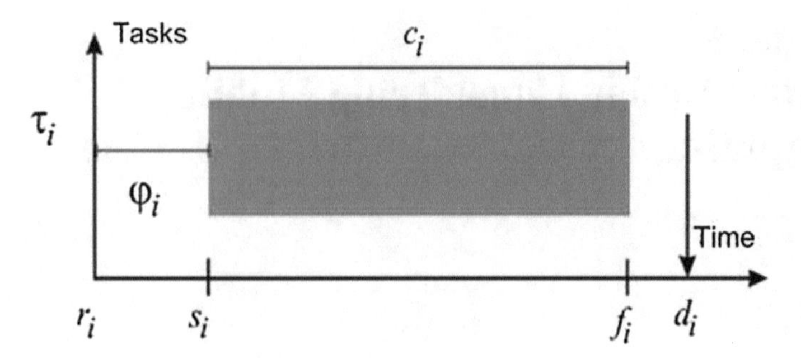

Fig. 5.1 Time parameters of a common task

a more detailed analysis of all real-time aspects that influence the activities of the system.

A real-time application is featured by a set Γ of activities, known as tasks, associated with certain time restrictions that have to be accomplished to achieve a wanted behaviour. It is frequent that the n tasks in real-time $\{\tau_0, \tau_1, \tau_2, \ldots, \tau_n\}$ execute after some time. From this, a task is understood as a sequence of identical activations $a_i, i = 1, \ldots, n$, separated between each other by a time-lapse [3].

The parameters that describe a real-time task are [3, 4]:

1. Arrival time (r_i) is the time in which a task is set for execution. It is also referred as release time.
2. Consumption time (c_i) is the time that takes to each task to execute its assigned activities without interruption. If it is a hard real-time system, it is considered as the Worst-Case Execution Time (WCET).
3. Delivery deadline (d_i) is the limit time in which a task should have finished. If the task is over before or at such an instant, the task is schedulable. If All tasks finish before all their delivery deadlines, the system is schedulable.
4. Start time (s_i) is the time in which a task starts its execution.
5. Finishing time (f_i) is the time in which a task finishes. It is also known as the response time. A task is schedulable if $f_i \leq d_i$.
6. Shifting (φ_i). Occasionally, a task may have a lag or shift since it does not necessarily start exactly at its initial time. Commonly, the start time of the first periodic instance is also known as shift.

Time parameters of a common task are shown in Fig. 5.1.

Reference [3] considers some further additional time parameters of tasks:

1. Criticality is a parameter related to the consequences of not accomplishing a deadline.
2. Value (v_i) represents the relative importance of a task regarding other tasks. It is not shown in the Fig. 5.1
3. Delay (l_i) represents the time delay of a task for achieving its objective regarding its deadline $l_i = f_i - d_i$.

4. Laxitude (x_i) is the maximum time that a task may delay its activation to achieve its deadline.

If all tasks have a shift equal to zero ($\varphi_i = 0, \forall i = 1, \ldots, n$), this is, all tasks start at the release time, the system is known to be synchronous; otherwise, the system is asynchronous.

There are other classifications for the real-time tasks that depend on the feature taken into consideration [3, 4]:

In terms of the way of execution, tasks may be preemptive or non-preemptive. A preemptive task is that whose execution can be interrupted by other tasks, to be continued later. A non-preemptive task has to be executed until completion of its consumption time, without interruptions.

In terms of the consequences produced by fault, tasks may be critical, if the fault causes catastrophic consequences, or uncritical, if the fault does not cause a serious consequence. Uncritical tasks are commonly used to deal with auxiliary functions, such as system maintenance, non-critical variable monitoring, etc.

In terms of their time characteristics, tasks may be:

1. Periodical, if their execution should repeat at a constant time interval, known as period (p_i). A periodical task is a sequence of identical invocations activated at a constant rate. A periodical task is represented as τ_i. Reference [1] describes a periodic task as a sequence of activations with the same parameters. This kind of tasks are characterized by time parameters $(c_i, d_i, p_i, \varphi_i)$ (Fig. 5.1). The kth activation of a periodic task τ_i is represented as a_{ik}, and it starts at the instant $\varphi_i + (k-1)p_i$, and should finish before the instant $\varphi_i + (k-1)p_i + d_i$.
2. Aperiodical, if they can be active at non-predictable time instants. For example, aperiodic tasks may be events generated by alarms. Aperiodic tasks can be classified as:

   - Without a deadline, which is non-critical, since they have no deadline to accomplish for their finishing. They can arrive at any moment, and they are only defined by their consumption time c_i.
   - With a hard deadline, they have a non-critical execution deadline. They are characterized by their consumption time and their deadline (c_i, d_i).

3. Sporadical, if a considerable fault is produced when they do not finish before their deadline. They are defined with the same parameters of a periodical task (c_i, d_i, p_i), taking into consideration that the period indicates the minimal time between two consecutive activations.

Scheduling

Basic references about scheduling and scheduling algorithms have been proposed during recent years, regarding their analysis, by [1, 3, 5]. In these, several aspects are used to identify the necessary considerations in the scheduling analysis of the reconfiguration strategies developed later in this book.

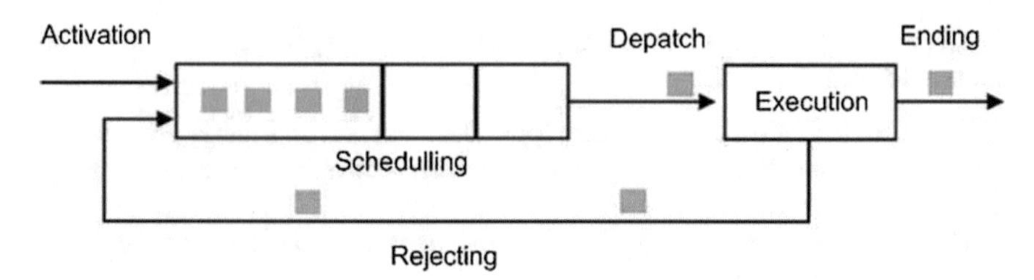

Fig. 5.2 A queue of tasks waiting for execution

A distinctive feature of real-time operating systems is the management of task execution, known as scheduling. For [3], defining the scheduling problem requires to specify three sets: a set of n tasks $\Gamma = \{\tau_0, \tau_1, \tau_2, \ldots, \tau_n\}$, a set of m processors $\Pi = \{\pi_1, \pi_2, \ldots, \pi_m\}$, and a set of s types of resources $\Phi = \{\phi_1, \phi_2, \ldots, \phi_m\}$. The precedence between tasks can be specified through an acyclic directed graph, and time restrictions can be associated with each task [3]. In this context, scheduling means to assign processors in Π and resources in Φ to tasks in Γ, aiming to complete all tasks under some imposed restrictions. This problem, in its general form, is NP-complete, and thus, intractable from a computing perspective. Certainly, the complexity of the scheduling algorithms is very relevant in real-time dynamic systems, in which scheduling decisions have to be taken on-line precisely during the execution of tasks.

In particular, when a single processor has to execute a set of concurrent tasks, the processor has to be assigned to different tasks, regarding a predefined criterion, namely the scheduling policy. The set of rules that at any moment determine the order in which tasks execute is known as a scheduling algorithm. The specific operation of the scheduling algorithm of assigning a processor to a particular task is known as dispatch. Hence, a task that could be executed on the processor can be (a) executing, if it has been selected by the scheduling algorithm, or (b) waiting for the processor, if there is another task in execution [1]. Normally, if tasks wait for the processor, they can commonly form a queue for execution. Figure 5.2 shows a queue of tasks waiting for execution [3].

References [3, 6] refer to tasks that potentially can be executed on the processor, independently of its availability, as an active task. A task waiting for the processor is an available task, while a task is currently executed on the processor is a task in execution. All available tasks, waiting for the processor, are kept in a queue known as the available queue. Operating systems that manage several kinds of tasks can have more than an available queue. In many operating systems that allow dynamic activation of tasks, a task in execution can be interrupted at any moment, so a more important task is able to immediately make use of the processor without any waiting in the available queue. In this case, the task in execution is interrupted and placed in the available queue while the processor is assigned to the more important task just at its arrival. The action of suspending a task in execution and place it in the available queue is known as pre-emption. Figure 5.3 shows the states of a process or task [3, 6].

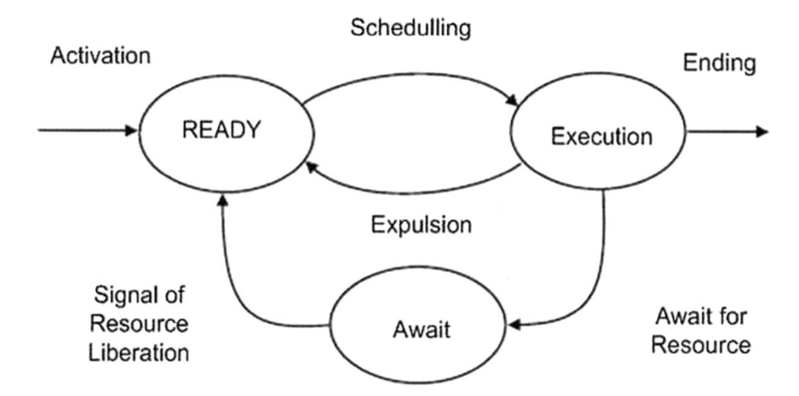

Fig. 5.3 States of a process or task

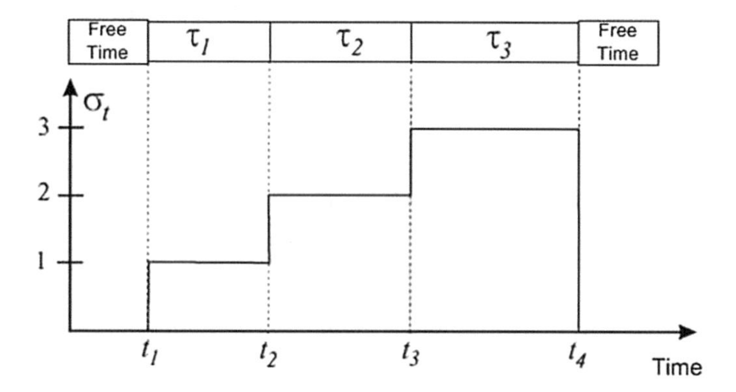

Fig. 5.4 An example of scheduling a set of tasks

Given a set of tasks $\Gamma = \{\tau_0, \tau_1, \ldots, \tau_n\}$, executing on a single processor, scheduling refers to assigning tasks to the processor, so each task is executed until it finishes. More formally, [3] defines scheduling as a function $\sigma = \Re \rightarrow N$ such that $\forall t \in \Re, \exists t_1, t_2$, and thus, $t \in [t_1, t_2)$ and $\forall t \in [t_1, t_2), \sigma(t) = \sigma(t)$. In other words, $\sigma(t)$ is an integer step function, and $\sigma(t) = k$, with $k > 0$, means that task τ_k is executing at time t, while $\sigma(t) = 0$ means that the processor is waiting.

Figure 5.4 shows an example of scheduling when executing a set of tasks $\Gamma = \{\tau_1, \tau_2, \tau_3\}$ on a processor [3]:

- For times t_1, t_2, t_3, and t_4, the processor performs a context switch.
- Each interval $[t_i, t_{i+1})$ in which $\sigma(t)$ is constant, refers to an amount of time in which task τ_i is executing. The interval $[x, y)$ identifies all values of t such that $x \leq t < y$.

A preemptive scheduling is that in which executing tasks can be arbitrarily suspended at any moment to assign the processor to another task, regarding a predefined

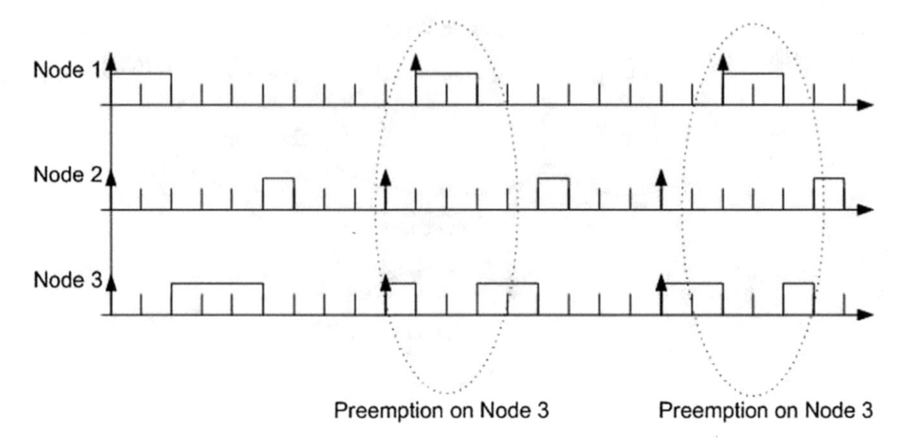

Fig. 5.5 A preemptive scheduling of the set of tasks Γ between node 3 and node 1

scheduling policy. In this kind of scheduling, tasks can be executed at disjoint time intervals.

A scheduling is called feasible if all tasks can be completed regarding a set of specific restrictions. Thus, a set of tasks is schedulable if there exists at least one algorithm which can produce a feasible scheduling.

A preemptive scheduling of the same example set of tasks Γ is shown in Fig. 5.5 [3].

Classification of Scheduling Algorithms

References [1, 3–5] reduce the complexity of implementing a scheduler by simplifying the architecture of the system. They choose a model preemptive or not, using pre-established execution priorities, eliminating precedence and/or limitations of the resources, assuming the activation of simultaneous tasks, sets of homogeneous tasks, this is, exclusively using periodic or aperiodic tasks. All these different features are used to classify several types of scheduling algorithms, proposed by the cited sources.

Among a great variety of scheduling algorithms for real-time tasks, there are the following:

- Preemptive. With preemptive algorithms, executing tasks can be interrupted at any moment to assign the processor to another active task, regarding a predefined scheduling policy.
- Non-preemptive. With non-preemptive algorithms, a task is executed on the processor until finishing it. In this case, all scheduling decisions are taken when a task finishes its execution.
- Static. Static algorithms are those that base their scheduling decisions on fixed parameters, assigned to tasks before their activation.
- Dynamic. Dynamic algorithms base their scheduling decisions on dynamic parameters, able to change during the evolution of the system.

- Off-line. A scheduling algorithm is used off-line if it is executed over the total set of tasks before their activation. The resulting scheduling is stored in a table and later executed by the dispatcher.
- On-line. A scheduling algorithm is said to be used online if the scheduling decisions are taken during the execution time, each time a new task enters the system, or when a task finishes its execution.
- Optimal. A scheduling algorithm is named optimal if it minimises some cost function defined over the set of tasks. When there is no such a function, and the only concern is achieving a feasible scheduling, then the algorithm is called feasible (suboptimal) if when using such an algorithm, it always finds the same scheduling.
- Heuristic. An algorithm is named heuristic if it tends to find an optimal scheduling, but it does not guarantee it for all cases.

More recent classifications of scheduling algorithms are gathered mainly in [7]:

- A static scheduler (off-line) requires the complete information about the scheduling problem at hand (number of tasks, deadlines, priorities, periods, etc.) to be known priori; however, with an online implementation, the configuration changes during execution, and thus, the scheduling problem has to be solved before the scheduler executes it.
- A scheduler is dynamic (on-line) if during execution time it is feasible to know the scheduling parameters, as well as performing the proposed configuration changes.

Gambier [7] highlights the advantage of determinism of static schedulers and as a liability their lack of flexibility. On the contrary, dynamic scheduling is very flexible, but it lacks certainty; on-line scheduling may have a poor performance if the system experiments an overload.

Consult [3, 4, 7] for a classification of dynamic schedulers. In this kind of schedulers, there is a family of schedulers called by priorities [3], which for each task, an important value is associated, known as priority. So, always the active task with the highest priority is executed. Priority schedulers can subdivide into the fixed priority schedulers and dynamic schedulers. The fixed priority schedulers, which assign an initial, immutable value of priority to each task, so each task keeps its priority during the whole execution. On the contrary, if the priority of each task changes during execution, then the scheduler is said to be of dynamic priorities.

Within the fixed priorities schedulers, the most common are the Rate Monotonic Scheduler (RMS) [8], which assigns the highest priority to the task with the shortest period, and the Deadline Monotonic Scheduler (DMS), in which the task with the highest priority is that with the closest deadline. Regarding dynamic priorities, the most widely studied algorithms are Earlier Deadline First (EDF) [8], which prioritises according to the absolute deadline, and Least Laxity First (LLF), which assigns the highest priority to the task with the least laxity.

Figure 5.6 presents a summary of the classification for scheduling algorithms available [7].

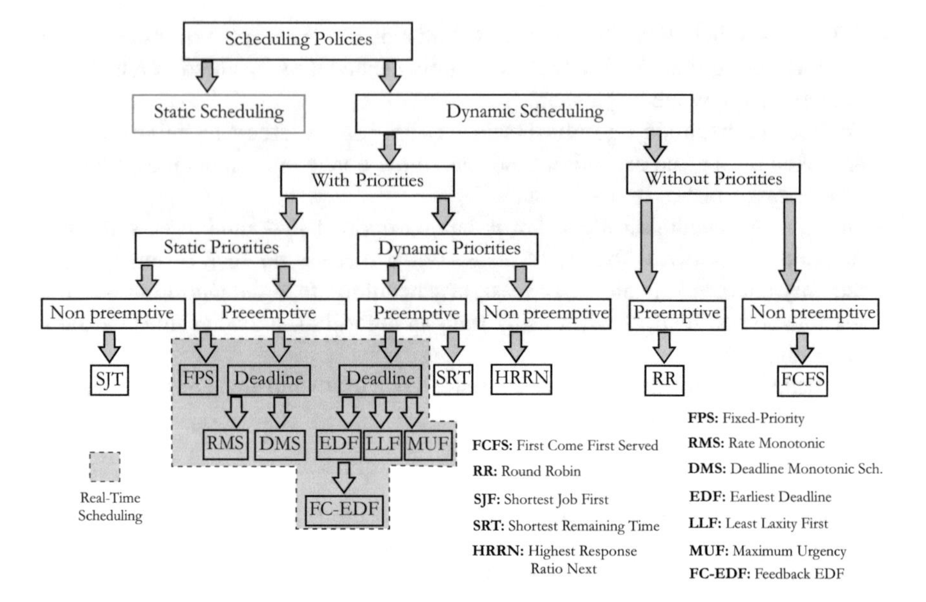

Fig. 5.6 Summary of the classification for scheduling algorithms

Scheduling of Periodic Tasks
In particular, given that in several real-time applications, the activities with the largest demand are constituted by periodic activities, here it is presented a review analysis of schedulability for the most well-known scheduling algorithms of periodic tasks. The schedulability analysis is presented for each algorithm, aiming to derive a guaranteed test for a set of tasks. According to with [1, 3–5], it is important at the beginning to define the used notation and develop the concept of utilisation factor.

The notation to represent the parameters of the task for the schedulability analysis [3] is shown in Table 2.1.

To simplify the schedulability analysis, the following assumptions about the tasks are made [1, 3]:

1. All instances of a periodic task τ_i are activated with a constant frequency. The interval p_i between two consecutive activations is the period of the task.
2. All the instances of a periodic task τ_i have the same worst execution time c_i.
3. All instances of a periodic task τ_i have the same relative deadline d_i, which is equal or less to their period p_i.
4. All tasks in Γ are independent, this is, there is no precedence relation or resource restriction among them.

Hence, in the cases of the previous assumptions, a periodic task τ_i can be completely characterized by its shift φ_i, the period p_i (as mentioned before), and the worst execution time c_i, and presented as [3]:

$$\Gamma = \{\tau_i\,(\varphi_i, p_i, c_i)\} \qquad i = 1, \ldots, n \tag{5.1}$$

The release time $r_{i,k}$ and the absolute deadline $D_{i,k}$ of the kth instance can be obtained as:

$$r_{i,k} = \varphi_i + (k-1)\,p_i \tag{5.2}$$

$$D_{i,k} = r_{i,k} + p_i = \varphi_i + kp_i \tag{5.3}$$

In such a context, it is considered that a task τ_i is feasible if all its activations finish before their deadlines. A set of tasks Γ is schedulable (feasible) if all the tasks within the set are feasible.

Hyperperiod. The concept of hyperperiod is defined as the interval with length Λ from which the pattern of activations of the tasks is repeated for a set of periodic tasks Γ. It is useful to ensure a scheduling interval as the representation horizon [4]. The maximum number M of activations for n tasks in each hyperperiod is equal to:

$$M = \sum_{i=1}^{n} \frac{\Lambda}{p_i} \tag{5.4}$$

The hyperperiod is the minimum common multiple of the periods of all tasks in Γ [1].

Utilization Factor. Reference [8] propose a utilization factor of a processor. Given a set Γ of n periodic tasks, the utilization factor of the processor is the time fraction that the processor spends in the execution of the set of tasks Γ. Given that c_i/p_i is the time fraction that the processor spends executing a task τ_i, the utilization factor for the n tasks is given as the summation of all the relations of this kind for each task, and bounded as follows:

$$U = \sum_{i=1}^{n} \frac{c_i}{p_i} \leq 1 \tag{5.5}$$

The utilisation factor of the processor provides a measure of the computational load of the processor due to a set of periodic tasks. Although the utilisation of the processor can be improved by incrementing the execution times of the tasks or decreasing their periods, there is a maximum value U under which Γ is schedulable, and above which Γ is not schedulable. Such a limit depends on the set of tasks, the particular relationships between periods of tasks, and the algorithm used to schedule the tasks.

Let $U_x(\Gamma, A)$ be the superior bound of the utilisation factor of the processor for a set of tasks Γ using algorithm A. When $U = U_x(\Gamma, A)$, it is considered that the set of tasks can be executed by the processor. In such a situation, Γ is schedulable by A, but any increment in the execution time of any of the tasks would make the set

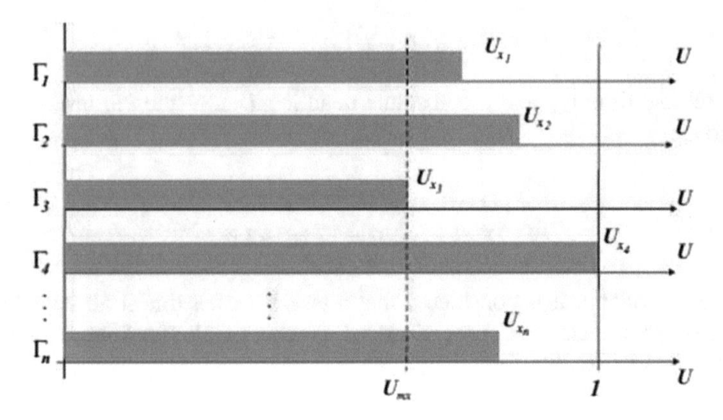

Fig. 5.7 $U_{mx}(A)$ for some scheduling algorithm A

Γ infeasible. For the given algorithm A, the minimal superior bound $U_{mx}(A)$ of the utilisation factor of the processor is the minimum of the utilisation factors of all the sets of tasks that completely make use of the processor:

$$U_{mx}(A) = \min_{\Gamma} U_x(\Gamma, A) \tag{5.6}$$

Figure 5.7 shows the meaning of $U_{mx}(A)$ for some scheduling algorithm A. The sets of tasks Γ_i in the figure differ in the number of tasks and in the value of their periods. When scheduled by algorithm A, each set of tasks Γ_i totally uses the processor when the utilisation factor U_i (changed when modifying the execution times of their tasks) reaches the particular superior bound U_{x_i}. If $U_i \leq U_{x_i}$, then Γ_i is schedulable; otherwise Γ_i is not schedulable.

Reference [3] points out that each set of tasks is able to have a different maximum bound. Given that $U_{mx}(A)$ is the minimum of all the maximum bounds, any set of tasks that has a utilisation factor of the processor under $U_{mx}(A)$ is schedulable by A.

$U_x(A)$ defines an important characteristic of the scheduling algorithms since it allows to easily verify the schedulability if a set of tasks. Certainly, any set of tasks whose utilisation factor is under this bound is schedulable by the algorithm A. In contrast, any utilisation factor above this bound would work if and only if the periods of the tasks are adequately related [1, 3].

If the utilization factor of the processor of a set of tasks is greater than one, the set cannot be schedulable by any algorithm. It is simple to show this, considering standardized periods and consumption times to integer time units. Let P be the product of all periods, $P = p_1 p_2 \ldots p_n$; if $U > 1$, then $UP > P$, which can be written as:

$$\sum_{i=1}^{n} \frac{P}{p_i} c_i > P \tag{5.7}$$

The factor P/p_i represents the number of times that τ_i is executed during the interval P, while Pc_i/p_i is the total computing time required by τ_i in the interval P. Therefore, the sum of the left side of the inequality in Eq. 5.7 represents the total computing time demand required by all tasks in P. Clearly, if such a demand exceeds the available time of the processor, there is no feasible scheduling for the set of tasks [3, 5].

A similar situation is presented in several distributed systems, regardless the communication media, like for example, wireless technology. In this sense, local networks provide an approachable situation like MANET [9].

Related Work. Several technologies have been previously combined to develop a protocol for MANET networks, allowing it to be context-aware. Such a context-aware feature is what allows triggering a distributed task every time an event occurs. In all previous work, the first step to execute a distributed task is to discover which nodes can perform the processes that compose such a distributed task.

This service (or process) discovery, regarding a mobile network, has been an important research topic during the last few years. Examples of these are the SLP-Manet [10] protocol, which modifies the Service Location Protocol [11] to work in MANET environments. Basically, this protocol establishes that service request packages are broadcasted, while service replies packages are cached by every node. However, due to this protocol is developed at the level of the application layer, it simply cannot access any routeing and scheduling information. Another problem is that cache entries may be false when the network topology changes.

Here, a discovery-less architecture is proposed, similar to [12]. Nevertheless, instead of AODV, DSR [13] is used here. Moreover, the protocols proposed by some related work require that only one service is discovered per message. Here, in order to execute a coordinated task, several processes should be found, and hence, the protocol proposed uses multiple service requests, encapsulated into one DSR route request.

Other approaches have been developed around the study of MANET networks and mobile networks, related to the topic here, such as those presented in [14–16]. These types of study have focused the research on context awareness, instead of process management. Further, the Coordinated Task protocol attempts to keep a low network load by merging all service discovery replies that belong to a particular Coordinated Task. Here, service discovery replies are piggy-backed on DSR route reply packets.

After process discovery, the related Coordinated Task should be executed. However, execute a distributed task on MANET is a challenge because during execution the availability of nodes may change: some mobile nodes become unavailable (they leave the network, or fail). Thus, they have to be replaced by other available nodes.

In [17], a distributed task is modelled as a Task Graph. An on-line algorithm is used to replace the unavailable nodes. However, this does not take into consideration the performance (in terms of execution time) or the network load, which makes this approach not suitable for the problem posed here, as RTDS.

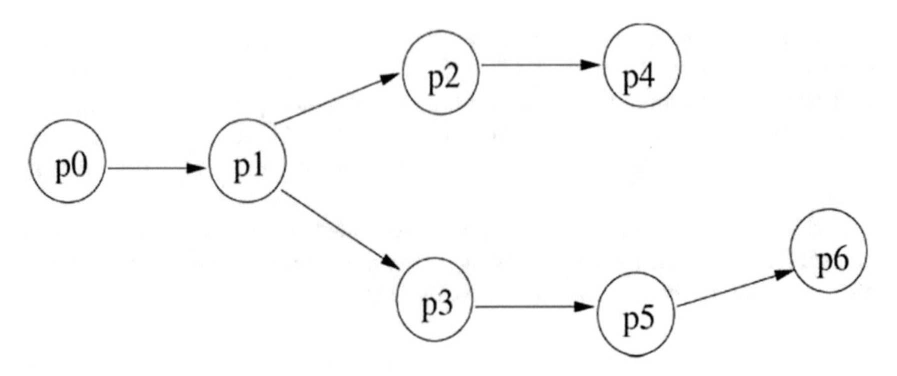

Fig. 5.8 Example of a task graph

Here, a similar Task Graph approach to [18] is used, but only to represent the data flow among of the processes. However, the network load is considered in order to map processes to nodes. Moreover, local cache information is used for replacing unavailable nodes, instead of performing new service requests.

Coordinated Tasks. A Coordinated Task is specified by using a control-driven coordination model [3, 19]. In this model, every process of the task has two ports: input and output. These ports are the only means of interaction between a process and its environment. Input and output ports from different processes are linked to form a communication flow. This communication model guarantees that the activity of each process is decoupled from the activity of any other process: each process is only in charge of performing a sequence of instructions that consumes and produces data. Data transmission between two nodes is the responsibility of the operating system on those nodes.

Executing a Coordinated Task creates a data flow in the network. This flow, called CT execution flow, is modelled using a Task Graph. The CT execution flow has a root node (represented as $p0$ in Fig. 5.8). This node contains the process in charge to detect the event that triggers the execution of the Coordinated Task.

The CT execution flow is implemented by using an adjacency list. Entries of such a list are tuples, composed of two process ids: one for the source process of the flow, and the other for the destination process. Tuples in the list are ordered according to the flow direction. So, every process may appear as a source or as a destination. For the CT execution flow in Fig. 5.8, the correspondent adjacency list is $(p_0, p_1), (p_1, p_2), (p_1, p_3), (p_2, p_4), (p_3, p_5), (p_5, p_6)$.

Now, the Coordination Task protocol is in charge of mapping processes of the Coordinated Task onto available nodes. This protocol also links the input ports of each process to the corresponding output ports. For the actual purposes, nodes with sensors are specifically used to start a Coordinated Task, and thus, they have a CT execution script per each event. Hence, every Coordinated Task has an associated execution script.

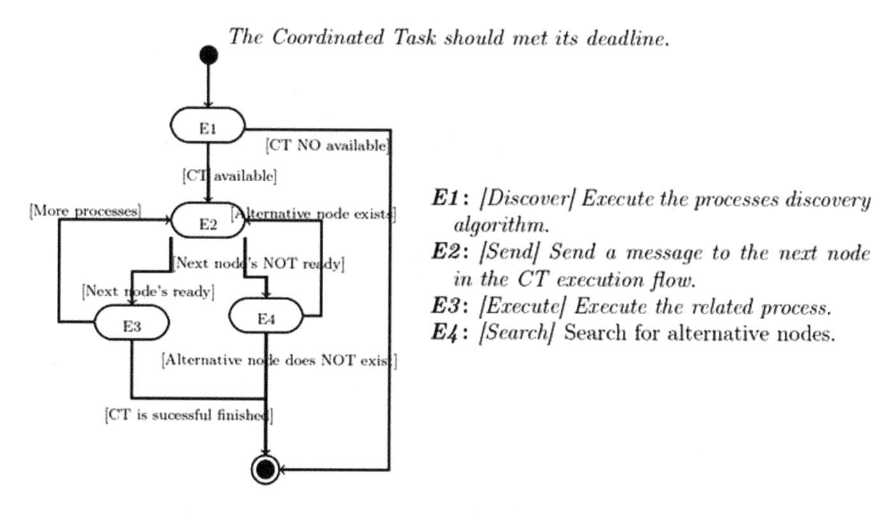

Fig. 5.9 Algorithm for executing a coordinated task

Commonly, the CT execution script includes the following information:

- Coordinated Task Name. A unique name that identifies a Coordinated Task.
- Trigger Event Identifier. A unique id that represents an event.
- Processes Identifiers. A field used to identify a process in a Coordinated Task. Depending on the application, the process id could be represented as simple as integers, or as complex as full XML descriptions, like WSDL.
- Processes Adjacency List. The adjacency list, describing the CT execution flow.
- Coordinated Task Soft Deadline. Every Coordinated Task should be completed before a deadline, and thus, this field represents such a deadline as a timeout value.
- Coordinated Task Maximum Communication Requirement. A list indicating how many data is transmitted by every process for each node. If a node does not produce data, its entry is the size of a CT Message.

Coordinated Task Execution Algorithm. Figure 5.9 shows the algorithm used to execute a Coordinated Task. Step E1 uses a reactive process discovery algorithm, based on encapsulating a discovery message into the DSR routeing protocol [20].

Responses to the discovery messages include the process worst execution time and the amount of data produced. These values are used in Eq. 5.8 to obtain the coordinated task execution time E_{CT}.

$$E_{CT} = T_D + T_T + T_E \tag{5.8}$$

where T_D is the time used by the process discovery procedure, and it is obtained by the node which detects the event. This node is also in charge of starting the process discovery algorithm.

On the other hand, T_T is the time needed to transmit the data. To determine T_T, it is required to find out the actual network capacity. However, how to do this is still an open issue. Here, an approach similar to that in [9] is followed. So, using the results, the transmission time (T_T) is calculated by using Eq. 5.9. Notice that, Eq. 5.9 considers that no message is sent in parallel.

$$T_T = \sum_{i=1}^{k} \left(\frac{\sqrt{n}}{W} L_i \right) \tag{5.9}$$

where k are the processes in the coordinated task, n is the number of nodes available in the network, W is the nominal network bandwidth, in kbytes/second, and L_i is the message length in kbytes. This value is obtained from the response to the discovery message.

Finally, T_E is the time used by the nodes to execute its assigned processes. It is calculated by simply adding all the worst execution time of each process. These worst execution values are transmitted as a field of the response to the discovery message. Each node obtains these values as presented in [9].

A Coordinated Task is considered feasible if its execution time (E_{CT}) satisfies Eq. 5.10, where D_{CT} represents the CT deadline.

$$E_{CT} \leq D_{CT} \tag{5.10}$$

The following step in the execution flow is started by sending a message to the node (step E2). If there is no MAC ACK message from this node, it is considered that the node is not available, and a search for an alternative node is started (step E4). On the contrary, if the node is available, it executes its processes (step E3). This procedure repeats until the Coordinated Task finishes, or until it is stopped due to the unavailability of any of the nodes required to perform the processes of this particular Coordinated Task. The scenario presented here is the worst case scenario, and therefore, time values are the extreme limits of this bounded process.

Scheduling Analysis

As stated before, commonly, periodic tasks represent the major demand for real-time systems. The need for periodic tasks arises from data acquisition by sensors, control loops, and monitoring systems, among others. Such activities need to be executed cyclically, at specific rates, derived from the requirements of the application. When the application consists of several concurrent tasks with individual time restrictions, the operating system must ensure that each periodic instance is regularly activated at its own rate and be completed before its deadline. For making a schedulable activation, it has to accomplish that the runtime of the task is less or equal to its deadline. In general, if all activations finish before or at the deadline, it is said that the real-time system is schedulable [3].

Feasibility and Optimality. As mentioned before, a valid scheduling is schedulable if each of its tasks is complete before its deadline. It is said that a set of tasks is schedulable regarding a scheduling algorithm if it uses such an algorithm, always producing a feasible scheduling [1]. The most used criterion to measure the performance of a scheduling algorithm is its ability to find a feasible scheduling for a set of tasks, always if such a scheduling exists. A scheduling algorithm is optimal if it always produces a feasible scheduling or what is the same if the set of tasks is feasible to be scheduled. On the contrary, if the optimal algorithm fails to find a feasible scheduling, it can be concluded that the set of tasks cannot be scheduled by another algorithm. Additionally to the criteria based on feasibility, other performance measures include the average value and the maximum finalisation value of a task regarding its deadline (lateness), how much it takes (tardiness), the response time, and the miss rate [1].

Most preemptive scheduling algorithms are priority driven. In this kind of scheduling, each task has associated a priority value, so the active task with the highest priority is always executed. These schedulers are sub-divided into schedulers with fixed priority and schedulers with dynamic priority. Fixed priority schedulers assign an initial priority value to each task, which is kept constant during the whole execution. On the contrary, if the priority of the task changes during execution, then the scheduler is said to be dynamic [4]. To carry out such a kind of scheduling, it is assumed that the time taken by the system to change context is negligible [21].

In particular, four basic algorithms for managing periodic tasks have been developed and extensively used: Rate Monotonic (RM) [8], Deadline Monotonic (DM) [22], Earliest Deadline First (EDF) [8], Least Laxity First (LLF) [2]; the first two make use of fixed priorities, while the other two use dynamic priorities.

In theory, schedulability is developed in two phases: the first phase, analysis, obtains as a result whether the set of tasks is schedulable or not by the selected algorithm; and the second phase, in which the tasks are scheduled. In the case of a dynamic algorithm, scheduling is carried out during execution time [4].

5.2 Consensus and Routing Algorithms

Here several approaches for selecting rates or taking decisions are presented since this definition is basic in order to configure proper mobile distributed systems.

Consensus Algorithms.

The consensus is one of the most widely studied problems in distributed systems. There are several proposals of algorithms for solving the consensus problem, which has been classified into regular, hierarchical, and random. They share some common characteristics, such as they are mainly composed of two operations: *proposal* and *decision*, as well as rounds. Each round is composed of both proposal and decision operations. In the proposal operation, a value is postulated to be broadcasted. In the decision operation, a function is applied each round to determine what value is chosen among all those proposals to be the result of that round. The result of the round may be a general consensus, or simply, a previous stage for another round [23].

Considering the notation previously used, a graph $G = (V; E)$ represents the network, in which V is the set of all nodes in the network, and E is the set of all edges. Let A be the set of all nodes that participate in the consensus, such that $A \subset V$. Each node proposes a link from itself to any of its neighbours at one hop. This is described using the notation as $propose\,(v, y, c)|\, v \in A$, in which y is a neighbor at one hop, and c is the cost associated to that link. For simplicity, this is expressed just as $P(c)$ [23].

Regular Consensus. This is also known as "flooding consensus". In this kind of consensus algorithm, each node proposes a value, representing which neighbour node y to communicate, in order to reach a certain destiny. The actual node broadcasts this information, so it is evaluated by the other nodes with which it has a connection. If any node disconnects at any moment, its information is still taken into consideration by the other nodes [23].

Hierarchical Consensus. In the hierarchical consensus, all nodes are ordered regarding certain priority. The node with the highest priority imposes its value to all the others, as rounds pass. No node can broadcast its value if its superior has not done so. If the highest priority node disconnects before broadcasting its value, then the value of the following node in the hierarchy is broadcasted. If this node does not fail, then its value is imposed on the rest of the nodes [23].

Random Consensus. This variant of consensus algorithms assures an integrity in the agreement. Each node has an initial value, which the node proposes through the primitive $P(c)$. This algorithm is carried out in rounds, each one composed of two phases. In the first phase, each node proposes a value. If when receiving the proposal of the neighbouring nodes it is detected a predominant value, this value is decided as the result, and it is proposed in the next phase of the round. However, if the node does not find a predominant value, then randomly selects a value among the initial values, and proposes it as the result of the next round.

Average Consensus. This variant of the consensus algorithms ensures convergence in the agreement. Each node proposes an initial value, using the primitive $P(c)$. The algorithm makes use of two rounds, in which *(a)* an initial value is proposed, and *(b)* a value is proposed as result, which is the average of the received values.

Routing Algorithms

In a wireless network, nodes may act as routers, allowing the communication of a source node with a destiny node by a series of point-to-point communications or hops from one routing node to the next. Hence, when a communication is requested, it has to be determined if a route exists that may connect the source node with the destiny node. Also, if a route is found, it has to be decided whether to use it or not. If there is no available route, it has to be decided what to do with the communication request.

Each one of these steps represents aspects of the routing problem. The selection of an available route constitutes such a routing problem, requiring that the available routes be defined. Such a definition may be the set of all the allowed routes given the structure of the network, or a convenient subset. This should include a description of the procedure to find routes and to select which route to use if there are more than one available routes. The decision to perform or not a communication through an available route is known as flow control [24].

An important feature of the communication network is that it should allow a route from a source node to a destiny node. If such a route does not change through time, it is called a static route, whereas if it does tend to change, it is called a dynamic route. In general, a routing algorithm is used to determine such a route [25]. There are two well-known routing algorithms for this, proposed by Bellman-Ford and by Dijkstra.

Types of Routing Algorithms. Routing algorithms indicate the route that a message has to take in order to reach its destiny. This is achieved in a wireless network by selecting the adequate output channel, which is selected from a set of possible options. routing algorithms can be classified into three types:

- Deterministic.
- Completely adaptive.
- Partially adaptive.

Routing Table. A routing table is the representation of the available routes within a communication network, from the perspective of a single node. The routing table of a node can be statically constructed, so the resulting routing table does not change through time, or it can be periodically (dynamically) updated, using information of the neighbouring nodes [26].

Adaptive Routing Algorithm. Dynamic routing algorithms allow to adapt to the traffic or topology changes of a network. A dynamic algorithm can be periodically executed, or directly as a response to the changes in the network. Nevertheless, adaptive routing algorithms are prone to route oscillations and loops.

5.3 Computer Network Design

Real-Time Design

A real-time (computing) system is that whose behaviour is fixed by the characteristics of its application [27]. Several characteristics have been identified for analysing and implementing real-time systems. Some of them are listed as follows:

1. Identify all the tasks that compose an application, as well as their time restrictions, such as deadline, period, activation time, etc., which have to be accomplished.
2. Specify the way in which to perform interruption management and context change.
3. Establish the way in which priorities are assigned to tasks.
4. Specify the way in which tasks are manipulated so they present certain kind of relation.
5. Perform the correspondent scheduling analysis.
6. Implement the proposed model.

In general, the design of real-time distributed systems covers the following characteristics [28]:

1. *Clock synchronisation.* Synchronisation is a more complex issue in distributed systems than it is in centralised systems since synchronisation in distributed systems requires the use of distributed algorithms. In practice, when there is a computer system or network with n nodes, the corresponding n clocks tend to oscillate at different rates, which implies an asynchronous execution. The difference between the time values is known as clock distortion δ.
2. *Event-driven systems and time-driven systems.* A real-time system is driven by events when the system is able to sense a significative event (using a sensor) and generates an interruption. On the other hand, a real-time system is time-driven when it is handled by a clock interruption each ΔT time units. This is known as clock mark. Detecting an event in this kind of real-time systems happens in the clock mark just next to the occurrence of the event. Thus, the selection of a value for ΔT should be carefully carried out: if it is too small, the real-time system would have a lot of clock interruptions, wasting time during the revision of the interruptions; if it is too large, crucial events could pass unnoticed.
3. *Predictability.* This is an inherent feature of real-time systems. It should be clear that the system accomplishes all its time limits, even during peak loads. The statistical analysis of behaviour, assuming independent events, show the frequent existence of errors, since there are common unexpected correlations between events. Random behaviour seldom occurs in a real-time system.

Desirable Properties of a Real-time System

- Timelines. This means that results not only have to be correct but also have to occur within time restrictions.
- Design for maximum load.
- Predictability.
- Fault tolerance.
- Maintenance.
- Continuous operation.

Basic Concepts
In computing, a process can be defined as the change of the state in the memory due to the action of the processor. As such, a process has an execution thread and can be composed of a finite set of tasks. A task is the sequential execution of code that does not suspend itself. Normally, tasks can be subdivided into smaller units of work, namely, jobs, execution instances, or subtasks. In general, the processor executes tasks in a certain order determined by a set of rules, known as a scheduling algorithm. This algorithm establishes that a task can exist in several states. A task is *executing* if the task has been chosen for execution or waiting to be assigned to the processor by the scheduling algorithm. A task waiting to be assigned is also known as a *ready to execute* task. An *active* task is that has been allocated on a ready to execute queue.

Several operating systems allow the dynamic activation of tasks. An executing task can be temporarily suspended at any moment, and hence, if a more important task requires execution, it can make use of the processor immediately. In this case, the executing task is interrupted and placed on the ready to execute queue, while the processor is assigned to the most important task in the ready to execute queue.

Restrictions
There are basically four types of restrictions to consider:

1. Restrictions imposed by the operating systems.
2. Time restrictions.
3. Precedence restrictions.
4. Mutual exclusion restrictions, due to shared resources.

Time Restrictions. This kind of restrictions refer to the constraints due to the very application at precise times. A typical time restriction of a real-time task is its deadline, which represents the time in which such a task has to complete its execution without provoking a damage to the system. If a deadline is specified regarding its activation time, then it is known as a relative deadline, while if it is defined regarding the initial time of the system (time zero), it is known as an absolute deadline.

A time restriction is strict if the user requires validation that the system always satisfies its time restrictions. In contrast, a time restriction is flexible if there is no need for such a constant validation, but only that the system statistically satisfies some time restrictions.

In general, a subtask J_i executing in real-time can be characterized by a subset of the following parameters (Fig. 5.10):

- Activation time r_i. It is the time in which a task is ready for execution; it is also referred as release time.
- Consumption time C_i.
- Period T_i.
- Absolut deadline d_i.
- Relative deadline D_i.

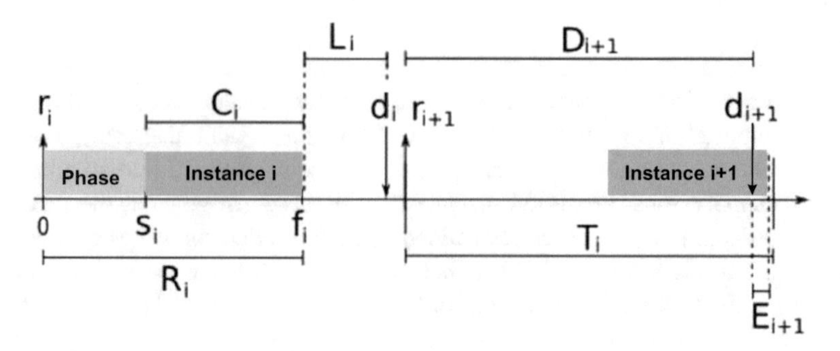

Fig. 5.10 Parameters of a subtask executing in real-time

- Start time s_i
- Final time f_i
- Response time R_i
- Latency L_i:
- Laxitud o excedente E_i:
- Phase ϕ_i:

Precedence Restrictions. Certain computing application require that activities are executed following a specified order, and thus, computing activities cannot be arbitrarily executed, but they have to respect some precedence relations defined during the design stage of the application. Such precedence relations are usually represented as an acyclic directed graph, in which tasks are represented as nodes and precedence relations by edges. The precedence in the graph induces a partial order of execution in the set of tasks.

Figure 5.11 shows an acyclic directed graph describing the relations of five tasks $(J_i \mid i = 1, \ldots, 5)$. From the graph structure, it can be noticed that task J_1 is the only task that can start execution since it has no predecessor. Finishing the execution of J_1, both J_2 and J_3 are able to start execution. Task J_4 can start execution only when J_2 has finished, while J_5 has to wait until both J_2 and J_3 finish their respective executions.

Mutual Exclusion Restrictions. This kind of restrictions is commonly due to concurrent execution of tasks, which share resources. Consistency problems may arise if two or more tasks make use of shared resources. To secure the consistency, concurrent tasks have to use access protocols, that commonly guarantee mutual exclusion when race conditions arise among concurrent tasks. Hence, these protocols assure a (mutually exclusive) sequential access to the shared resources. In general, most operating systems provide with synchronisation mechanisms that can be used for developing the access protocols.

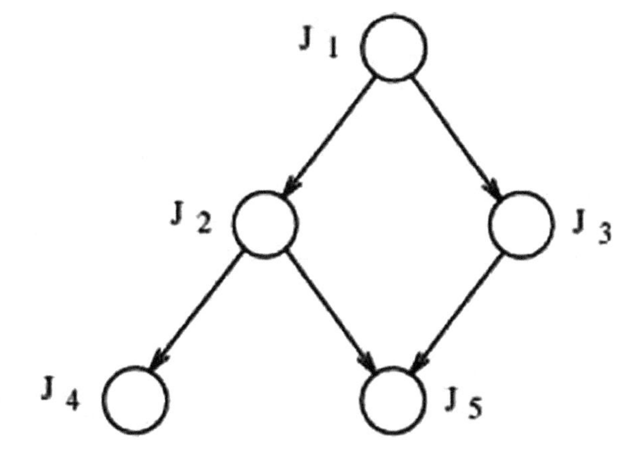

Fig. 5.11 An acyclic directed graph describing the relations of five tasks

The Scheduling Problem
In order to define the scheduling problem, it is necessary to define three sets:

1. A set of n tasks $\Gamma = \{J_1, \ldots, J_n\}$,
2. a set of m processors $P = \{P_1, \ldots, P_m\}$, and
3. a set of r kinds of resources $R = \{R_1, \ldots, R_r\}$.

Besides, there can be some specifications about precedence relations among tasks, as well as time restrictions associated with each task. Thus, scheduling means to assign resources R and processors P to the tasks in Γ, aiming to complete all the tasks under the imposed restrictions.

Off-line Scheduling Algorithms
Usually, the feasibility of scheduling is defined in terms of accomplishing time restrictions (deadlines, task dependencies, and so on). For the off-line scheduling algorithms, the solution search space can be represented as a search tree, in which each node represents the assignation of a task to a processor. Any branch to a leaf represents a schedule, although it is not necessarily feasible. A partial scheduling is a route between any pair of nodes in the tree.

Once assignation has been done by each node in the tree, the algorithm traverses the tree by means of a deep search, running all those branches that do not offer a feasible scheduling, i.e., do not accomplish with the specified restrictions. When the search finishes, each possible route represents a possible schedule. In case of which there is more than one feasible schedule, the objective is to find the optimal schedule, in which optimality is defined as a function that has to be maximised or minimised. Examples of these functions are the average load, the maximum workload, reducing execution time, reducing execution latency, etc.

On-line Scheduling Algorithms
Periodic Tasks Scheduling. The existence of periodic tasks in a computing system is due to the need to produce a response, acquire, and process data with certain reg-

ularity. Basically, the scheduling strategies distribute processor time so the temporal restrictions are satisfied.

Here, three periodic tasks scheduling techniques are reviewed: rate monotonic (RM), earliest deadline first (EDF), and deadline monotonic (DM). These algorithms are based on priority, which requires that the processor remains idle only when no executing task requires such resources. Algorithms based on priority can be static or dynamic. In a static algorithm, all tasks are divided into subsets. In contrast, in a dynamic algorithm, ready for execution tasks are placed on a queue, and are dispatched to processors for their execution as soon as a processor is made available.

These techniques are widely used in real-time systems, since using these algorithms it is possible to specify a criterion for the use of the available resources in a system. Scheduling algorithms allow the rational utilisation of resources, and occasionally, with the adequate conditions, they can reach optimal solutions. Unfortunately, it is simply not possible to establish rules that define when there are optimal conditions for implementing a specific approach, and totally take advantage of the resources.

Rate Monotonic. The RM algorithm [8] executes under the following assumptions:

1. Tasks are totally preemptive and, besides, the exchange between tasks has a negligible cost.
2. Only processing requirements are significative. Memory and I/O are negligible.
3. All tasks are independent, this is, there is no precedence relationship.
4. All tasks are periodic.
5. Relative deadlines of tasks are equal to their periods.

In this scheduling algorithm, priorities are fixed, since they are assigned based on the period: the task with the shortest period is the one with the highest priority. In other words, for a finite set of periodic tasks $\Gamma = \tau_i : i = 1, \ldots, N$, whose periods T_i accomplish that $T_1 < T_2 < \ldots < T_N$, the task with the identifier $i = 1$ is the one with the highest priority.

To verify if a set of tasks is schedulable using this algorithm, there is a schedulability test, which consists of verifying that the total utilization of tasks is not greater than $N\left(2^{1/N} - 1\right)$:

$$U = \sum_{i=1}^{N} \frac{C_i}{T_i} < N\left(2^{1/N} - 1\right) \leq 1 \qquad (5.11)$$

Deadline Monotonic. The DM algorithm [8] executes under the following assumptions:

1. Tasks are totally preemptive, and the cost of such a property is negligible.
2. Only processing requirements are significative. Memory, I/O, and other requirements are negligible.
3. All tasks are independent. There is no precedence relationship.

4. All tasks are periodic.
5. Periodic tasks do not have the same relative deadline.

Earliest Deadline First. The EDF algorithm [8] executes under the following assumptions:

1. All tasks τ_i have the same worst execution time C_i.
2. Tasks are totally preemptive, and the cost of this is negligible.
3. All tasks are independent. There is no precedence relationship.

Scheduling Tasks in Multiple Processors and Distributed Systems
Multiple processor systems require a communication media for the information exchange. Depending on characteristics of the memory architecture, there are two types of multiprocessor systems: *shared memory*, which implies the existence of a physical memory, shared among the processors of the system, and *distributed memory*, in which processors do not have a common memory, but communicate using I/O operations.

Shared memory multiprocessor systems are commonly tightly coupled systems, this is, the communication time between processor is very small when compared with the processing time of a task, and thus, can be ignored. Hence, communication is reduced to read and write operations over the shared memory. Distributed memory multiprocessor systems, in contrast, are loosely coupled, since the communication time between processors tends to be comparable to the processing time of tasks.

Real-time scheduling of distributed systems with multiple processors poses two subproblems: assigning tasks to processors, and scheduling of tasks on individual processors. The problem of assigning tasks refers to how to divide the set of tasks, and then, how to assign them to processors. The assigning tasks problem can be static or dynamic.

Static Assingning
Balancing Algorithm. This algorithm consists of a centralised data structure, i.e., a queue, which maintains tasks in incremental order, according to their utilisation. So, tasks are removed one by one from the data structure, each time sending them to the node with the least utility [29].

The objective function to minimize in this algorithm is $\sum_{i=1}^{n} |u - u_i|$, where u is the average utilization of the processors in the system, and u_i is the utilization value of processor i. This algorithm is factible when the number of processors in the system is odd.

Algorithm Next-Fit for Rate Monotonic. In this algorithm [8], a set of tasks is divided, so each subset of tasks is assigned to a processor that makes use of Rate Monotonic for scheduling. This algorithm attempts to use the least number of possible processors. This algorithm, when compared with the balancing algorithm, does not require that the set of processors be given beforehand or predetermined. This algorithm classifies different tasks depending on their utilisation. One or more processors are assigned to a certain kind of task. The essence of this algorithm is grouping tasks with a similar utilisation for their execution.

Tasks are classified regarding their utilization through the following policy: if tasks are to be divided into m classes, a task τ_i belongs to class k, such that $0 \leq k < m$, if and only if:

$$\left(2^{\frac{1}{k+1}} - 1\right) < \frac{C_i}{T_i} \leq \left(2^{\frac{1}{k}} - 1\right) \tag{5.12}$$

Buddy algorithm. This algorithm aims to overcome the high amount of communications of the auction algorithm [8]. In this algorithm, a processor can be in two states: overload or underload. A processor (P_i) has the status of underloaded if its utilization is not larger than a pre-established value Th, ($u_i < Th$). A processor is overloaded if this value is over such a pre-established value ($u_i \geq Th$). In this algorithm, a processor sends its information only when its status changes from overloaded to underloaded, or vice versa.

Scheduling of Coordinated Tasks

The Coordination Triggered Protocol (CTP) enables nodes of the system with the capacity to coordinate among themselves, aiming to carry out a coordinated task. Such a coordinated task is composed of several processes, distributed throughout the network. CTP performs the following main functions [19]:

1. Mapping processes to nodes.
2. Provide mechanisms that nodes require to perform the coordinated task.
3. Provide a communication infrastructure.
4. Monitor the execution status of a coordinated task.
5. Guarantee that at the end nodes release their used resources.

Consensus

Consensus is a procedure, this is, a finite number of steps whose purpose is reaching an agreement about some value. An important aspect of achieving consensus, applied for many practical situations, is that consensus should be achieved in the presence of faults. Achieving consensus in a distributed system without faults seems not to be a real challenge. It is enough to exchange messages with the needed information to achieve an agreement [30]. In contrast, when a decision is taken even in the presence of faults, or if the consensus interacts with a scheduling algorithm, the problem seems more interesting.

Agreements in the consensus not only allow establishing the value of some important variables in distributed systems. In practice, a consensus is very important to maintain consistency in some systems. Consensus has been studied based on different types of system models: for time models, as synchronous and asynchronous models; for fault models, under byzantine, crash, or due to communication faults. Regarding asynchronous systems, the consensus is restricted by the result produced by Fischer, Lynch, and Patterson (FLP), which assures that it is impossible to reach a deterministic consensus in an asynchronous system [31].

5.4 Control Design Using Local Mobile Conditions

Energy consumption for mobile nodes is an interesting and important issue in order to determine how confident is a node in the network and how stable maybe the system, however, a deep exploration in the issue is out of the scope of this book, interesting references have studied this issue as follows.

Routing Information Protocol (RIP)
Each node builds a unidimensional array (a vector) that contains the distances (costs) to all nodes and distributes such a vector to their immediate neighbours. Initially, it is assumed that each node is aware of the link cost to each one of its immediate neighbours. A fallen link is considered to have an infinite cost [32].

The routing protocols are different from the graph model previously described. In a group of networks, the aim of routers is to learn how to re-send packets to several networks. Thus, instead of considering the cost of reaching other routers, routers broadcast the costs for reaching a particular network.

RIP is the most direct implementation of routing by distance vector. A router that makes use of RIP sends calls every 30 s. It also sends calls when there is a change in its routing table. In practice, the protocol attempts to reach the target in less than 16 hops. In fact, 16 hops are equivalent to an infinite distance [32].

Open Shortest Path First (OSPF)
routing by link state is the second more used routing protocol. The initial considerations are the same as RIP. It is assumed that each node is able to detect the state of the links which communicate with immediate neighbours (connected or disconnected) and the cost of each link [32].

In practice, each router independently obtains its own routing table from the state packet of link 5. This is used to perform Dijkstra's Search Forward algorithm.

Optimized Link State Routing (OLSR)
OLSR is a proactive routing protocol that distributes jobs to establish connections between nodes in an ad-hoc wireless network. Flooding (direct broadcast of information throughout all the network) is inefficient and very costly for a mobile wireless network, due to bandwidth limitations and scarce channel quality. OLSR provides an efficient mechanism for information broadcast, based on Multipoint Relays (MPR) [33]. Using MPR, instead of allowing each node to re-transmit any received message (as stated in classic flooding), all nodes in the network select among their neighbours a set of multipoint relays. These are in charge of retransmitting the messages sent

by the actual node. The other neighbours cannot re-transmit. This reduces the traffic generated by a flooding [33].

Independently of the way for electing them, the set of MPRs of a node should verify that they are able to reach all the neighbours at a distance of two hops from the actual node. This is a covering criterion of the MPR [33].

Topology Control (TC) messages are periodically and asynchronously sent. Through these, nodes inform about the close network topology. In contrast with HELLO, TC messages are global and should reach to all nodes in the network.

Temporally-Ordered Routing Algorithm (TORA)

TORA is a distributed routing protocol for mobile wireless networks. This protocol aims for a high degree of scalability using a flat algorithm, this is a non-hierarchical routing algorithm. This algorithm attempts to suppress the propagation of control messages to the most far away levels of the network. TORA builds up and maintains a directed acyclic graph, with root in its destiny [34].

TORA is a reactive protocol, characterised by providing to the sender node, not only one, but multiple trajectories to arrive at its destination. The procedure is as follows: each node performs a copy of TORA for each destiny, and the protocol creates, maintains, and cancels the routing trajectories. TORA associates a weight to each node of the network, depending on its destiny. Thus, messages move from a node with larger weight to a node with less weight, while routes discovered by QRY (query) packets are updated with those of type UPD (Update) [34].

The node that responds makes use of the UPD packet, adding its weight. UPD packets are broadcasted, so this allows to all nodes in the middle to conveniently update their weights. From this, all nodes aiming to a far or unreachable destiny increase their local weight up until reaching an allowed maximum value, while a node that finds a close neighbour with a weight tending to infinity will change trajectory [34].

The TORA protocol supports the Internet MANET Encapsulation Protocol (IMEP), which provides a trustworthy expedition service for routing protocols. Adding into a unique block IMEP and TORA messages reduce the network overhead and test the state of neighbouring nodes. To obtain such an addition, it is periodically used the exchange of two messages, known as BEACON and HELLO [34].

Related Work

In an ad-hoc network, all nodes send a HELLO message to check their connectivity with their neighbouring nodes. In this scheme, each node keeps an information table about its neighbouring nodes. The agent node broadcasts a request to all neighbouring nodes, and these resend this request to their neighbours, and so on. When all neighbours are covered, a convercast message is sent with the node address, and its adjacency matrix considering all its neighbours. When this message is received, the agent node obtains the adjacency matrix of one hop for each node in all the network. It may be the case that a node receives more than two messages. In this case, the node checks the sequence number of the message, and stores it or updates the adjacency matrix with the larger sequence number [35].

Another important related work in this area refers to the routing algorithms that attempt to avoid or diminish the use of broadcast. This is the case of the Bypass Routing algorithm [36], which behaves as a recovery mechanism for routes and local errors. Essentially, this algorithm recovers from a failed route. First, the node searches for an alternative route in its route cache memory. If the route exists, the node patches the broken route with an alternative route.

In the context of consensus algorithms, one of the most outstanding related work regarding our case study is the voting algorithm using a weight average [37]. This algorithm proposes a novel way to incorporate consensus using the average of the received votes as initial values.

5.5 Concluding Remarks

the mobile computing is studied considering consensus and routing algorithms as the respective strategies for local agreements in order to provide a coherence in particular movements. This sort of strategy and integrated to scheduling approximation lead us to a more complex time delay behaviour that modifies the relationship of local sensor-controller and controller-actuator time delays as presented in the last part of this chapter.

References

1. Liu, F., Narayanan, A., Bai, Q.: Real-Time Systems (2000)
2. Mok, A. and Dertouzos, M.: Multiprocessor scheduling in a hard real time environment. Texas Comput. Syst. **17** (1978)
3. Buttazzo, G.C.: Real Time Systems. Springer (2004)
4. Balbastre, P.: Modelos de Tareas para la Integracion del Control y la Planificacion en Sistemas de Tiempo Real. Tesis Doctoral, Universidad Politecnica de Valencia (2002)
5. Cheng, A.M.: Real-Time Systems: Scheduling, Analysis, and Verification. Wiley (2002)
6. Tanenbaum, A.S.: Sistemas Operativos Modernos. Prentice Hall Hispanoamericana (1993)
7. Gambier, A.: Real-time control systems: a tutorial. In: 5th Asian Control Conference, 2004, vol. 2, pp. 1024–1031. IEEE (2004)
8. Liu, C., Layland, J.: Scheduling algorithms for multiprogramming in a hard-real-time environment. J. Assoc. Comput. Mach. **20**(1), 46–61 (1973)
9. Palomera-Perez, M., Benitez-Perez, H., Ortega-Arjona, J.: Coordinated tasks: a framework for distributed task in mobile area networks. Express Lett. ICIC **4**(5) (2011)
10. El Saoud, M.A., Konz, T., Mahmoud, S.: SLPManet: service location protocol for Manet. In: Procedings of the International Conference on Wireless Comunications on Mobile Computing IWCMC (2006)
11. Guttum, E.: Service location protocol: automatic discovery of IP network services. Internet Comput. IEEE (1999)
12. Engelstad, P.E., Zheng, Y., Koodli, R., Perkins, C.E.: Service discovery architectures for on-demand ad hoc networks. Ad Hoc Sens. Wirel. (2006)

13. Johnson, D., Hu, Y., Maltz, D.: The dynamic source routing protocol (DSR) for mobile ad hoc networks for IPv4. Technical Report Internet Draft (2003)
14. Basu, P., Ke, W., Little, T.D.C.: A novel approach for execution of distributed tasks on mobile ad hoc networks. Ad Hoc Netw. IEEE (2002)
15. Papadopoulos, G.A., Arbab, F.: Coordination Models and Languages, Centrum voor Wiskunde en Informatica (CWI) (1998)
16. Christensen, E., Curbera, F., Meredith, G., Weerawarana, S.: Web Service Definition Language Coordination Models and Languages (WSDL). http://www.w3.org/TR/wsdl
17. Engelstad, P.E., Zheng, Y.: Evaluation of service discovery architectures for mobile ad hoc networks. In: The Second Annual Conference on Wireless On-demand Network Systems and Services, pp. 2–15 (2005)
18. Gupta, P., Kumar, P.R.: The capacity of wireless networks. Trans. Inf. Theory **46–2**, 388–404 (2002)
19. Palomera-Perez, M., Benitez-Perez, H.: Scheduling coordinated task. In: Conference on Industrial Electronics and Application. IEEE (2009)
20. Chiang, M., Wong, S.: Hybrid consensous agreement on CDs based mobile ad hoc network. IJICIC **5–8**, 2291–2309 (2009)
21. Xia, F., Sun, Y.: Control and Scheduling Codesign: Flexible Resource Management in Real-Time Control Systems. Springer (2008)
22. Joseph, Y., Leung, T.: On the complexity of fixed-priority scheduling of periodic real-time tasks. Perform. Evolution **4**(2), 237–250 (1982)
23. Guerraoui, R., Rodrigues, L.: Introduction to Reliable Distributed Programming. Springer New York Inc., USA (2006)
24. Girard, A.: Routing and Dimensioning in Circuit-Switched Networks, 1st edn. Addison-Wesley Longman Publishing Co., Inc., Boston, MA, USA (1990)
25. Medhi, D., Ramasamy, K.: Network Routing: Algorithms, Protocols, and Architectures. Morgan Kaufmann Publishers Inc., San Francisco, CA, USA (2007)
26. Galvin, P., Gagne, G., Siberschartz, A.: Sistemas Operativos. Limusa Wiley (2002)
27. Hermosillo-Gomez, J.A.: Consenso en ambientes no homogeneos en base a una plataforma de tiempo real. Tesis para obtener el grado de Maestria en Ciencia e Ingenieria de la Computacion, UNAM (2013)
28. Fischer, M.J., Lynch, N., Paterson, M.S.: Impossibility of distributed consensus with one fauty process. J. Assoc. Comput. Mach. **32**, 374–382 (1985)
29. Esquivel-Flores, O.A.: Estudio de Sistemas Multi-agentes Reconfigurables. Posgrado en Ciencias e Ingenieria de la Computacion, UNAM. 23 Enero (2013)
30. Khanna, V.K., Singh, S.: An improved pigback ethernet protocol and its analysis. Comput. Netw. ISDN Syst. **26**(11), 1437–1446 (1994)
31. Mendez-Monroy, P.E., Benitez-Perez, H.: Supervisory fuzzy control for networked control systems. Int. J. Innov. Comput. Inf. Control Express Lett. **3–2**, 233–240 (2009)
32. Peterson, L.L., Davie, B.S.: Computer Networks: A Systems Approach, 4th edn. The Morgan Kaufmann Series in Networking. Morgan Kaufmann (2007)
33. Jacquet, P., Mühlethaler, P., Clausen, T., Laouiti, A., Qayyum, A., Viennot, L.: Optimized link state routing protocol for ad hoc networks, pp. 62–68 (2001)
34. Park, V.D., Corson, M.S.: A highly adaptive distributed routing algorithm for mobile wireless networks (1997)
35. Liu, Z., Kim, J., Lee, B., Kim, C.G.: A routing protocol based on adjacency matrix in ad hoc mobile networks. In: International Conference on Advanced Language Processing and Web Information Technology, pp. 430–436 (2008)
36. Sengul, C., Kravets, R.: Bypass routing: an on-demand local recovery protocol for ad hoc networks (2004)
37. Latif-Shabgahi, G., Bass, J.M., Bennett, S.: Histiry-based weighted average voter: a novel software voting algorithm for fault-tolerant computer systems. In: PDP, pp. 402–409. IEEE Computer Society (2001)

Chapter 6
Applications

Abstract In previous chapters, NCS modelling and control have been presented using control, scheduling and codesign strategies, in order to reduce the effects of including a communication network within the control loop. This chapter shows several applications of the proposed design strategies for NCS. The first section shows the two case studies, one is a SISO MAGLEV system and MIMO 2-DOF helicopter system, both are used in different NCS configuration into the chapter to prove the effectiveness of proposed methods. Then, the design of an adaptive fuzzy control for NCS (Sect. 4.2) is shown for both the SISO and MIMO system through the fuzzy systems, realizing the model of the time delay, and designing the model and controller of the system with both time delays and lost packets, where the MIMO system shows the versatility of the design with multiple outputs and a fully distributed configuration. The third section presents the application of sampling frequency control (Sect. 4.3) to the SISO and MIMO systems, showing an excellent system performance by controlling the frequency of transmission of the agents. Finally, the controller-scheduling codesign of Sect. 4.4 is applied to the MIMO system using two fuzzy models, a control model allows to generate a control law for multiple sampling periods based on the quality of control while a scheduling model selects a sampling period, both in function of the network and system performance. Showing that the combined application of strategies improves the performance of the complete system.

6.1 Case Studies

The MAGLEV system is a single input-single output (SISO), nonlinear, open-loop, unstable, and time-varying system. This system contains an electromagnet to levitate a steel ball, an infrared sensor for the measurement of the steel ball position and a current sensor in the coil. The aim is to design a controller to levitate the steel ball following a desired trajectory (Fig. 6.1).

The Magnetic Levitation equations are presented regarding the nonlinear model using the free body diagram:

H. Benítez-Pérez et al., *Control Strategies and Co-Design of Networked Control Systems*, Modeling and Optimization in Science and Technologies 13,
https://doi.org/10.1007/978-3-319-97044-8_6

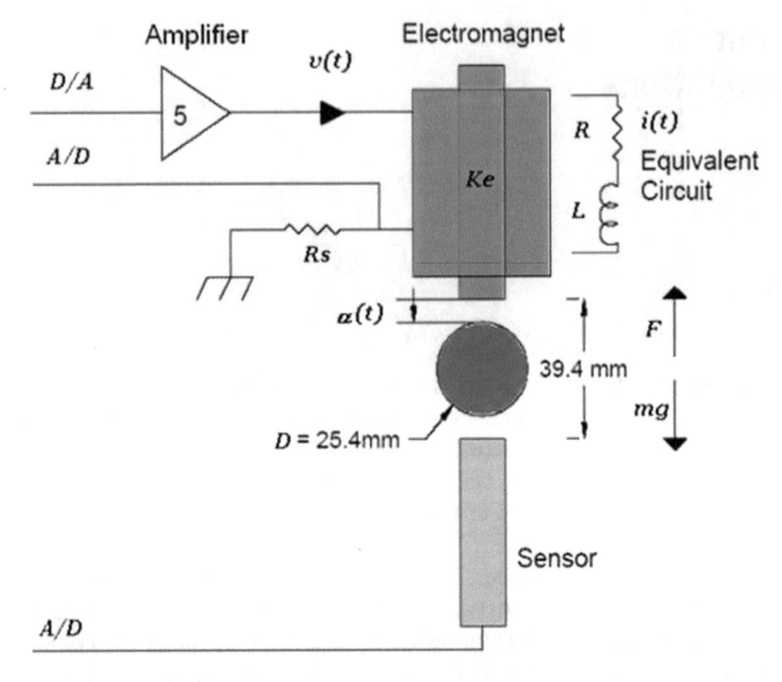

Fig. 6.1 Model of the levitation system

$$\dot{x_1} = x_2 \tag{6.1}$$

$$\dot{x_2} = \frac{-K_m x_3^2}{2M_b(x_1)^2} + g \tag{6.2}$$

$$\dot{x_3} = \frac{1}{L_c}(-Rx_3 + u) \tag{6.3}$$

where $R = R_c + R_s$ and $u = V_c$ is the input voltage, and:

- R_c is the electromagnet resistance.
- R_s is the resistor in series wiht the coil.
- K_m is the constant of electromagnet force.
- M_b is the mass of the ball.
- g is the gravitacional constant.
- L_c is the coil inductance.

The actual values of related variables are shown in Table 6.1.

2-DOF Helicopter

The case study is a multi input-multi output (MIMO), non-linear, open-loop unstable and time-varying system. The 2-DOF helicopter experiment consists of a helicopter model, mounted on a fixed base, with a total weight of 3.5 kg. It has two propellers driven by 12-volt DC motors, as shown in Fig. 6.2, that has been presented in Chap. 3.

Table 6.1 MAGLEV parameters

Symbol	Value	Unit
L_c	0.4125	H
R_c	10	Ω
R_s	1	Ω
k_m	$6.5308e^{-5}$	Nm^2/Amp^2
M_b	0.068	kg
g	9.81	m/s^2

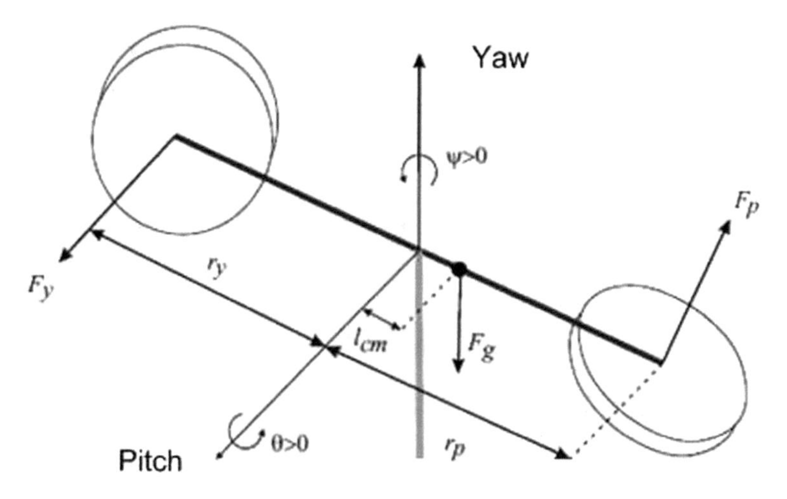

Fig. 6.2 Helicopter model

The front propeller controls the elevation of the helicopter nose about the pitch axis, with a range of motion ±45° from horizontal and the back propeller controls the lateral motions of the helicopter about the yaw axis.

The pitch and yaw angles are measured using high-resolution encoders, with 0.08° for the pitch angle, and 0.04° for the yaw angle. The encoders and motors signals are transmitted using a slip ring. This eliminates the possibility of wire tangling on the yaw axis and allows the yaw angle to rotate freely about 360°. The Euler-Lagrange method is used to derive the nonlinear equations, describing the motion of the helicopter (Eq. 6.4). From the non-linear equations of motion, the linear discrete state space models are obtained.

Table 6.2 System specifications

Symbol	Value	Unit
J_{eqp}	0.0348	kg· m^2
J_{eqy}	0.0432	kg· m^2
K_{pp}	0.204	N· m/V
K_{yy}	0.072	N· m/V
K_{py}	0.0068	N· m/V
K_{yp}	0.0219	N· m/V
B_{qp}	0.800	N/V
B_{qy}	0.318	N/V
m_{heli}	1.3872	kg
l_{heli}	0.186	m

$$\begin{aligned} &D(q)\ddot{q} + C(q,\dot{q})\dot{q} + g(q) = \alpha \\ &D(q) = \begin{bmatrix} J_p + ml^2 & 0 \\ 0 & J_y l^2 \cos\theta^2 \end{bmatrix} \\ &C(q,\dot{q}) = \begin{bmatrix} B_p & ml^2 \sin\theta\cos\theta\dot{\psi} \\ -2ml^2 \sin\theta\cos\theta\dot{\psi} & B_y \end{bmatrix} \\ &g(q) = \begin{bmatrix} mgl\cos\theta \\ 0 \end{bmatrix} \\ &\alpha = \begin{bmatrix} K_{pp}V_p + K_{py}V_y \\ K_{yp}V_p + K_{yy}V_y \end{bmatrix} \end{aligned} \tag{6.4}$$

where the variables are defined as:

- K_{pp} is the thrust torque acting on the pitch axis of the pitch motor
- K_{py} is the thrust torque acting on the yaw axis from the pitch motor
- K_{yp} is the thrust torque acting on pitch axis from the yaw motor
- K_{yy} is the thrust torque acting on the yaw axis from the yaw motor
- B_p is the equivalent viscous damping about the pitch axis
- B_y is the equivalent viscous damping about the yaw axis
- l_{heli} is the length from the centre of mass along the body of the helicopter
- g is the gravity acceleration
- m_{heli} is the total moving mass of the helicopter
- J_p is the total moment of inertia about the pitch pivot
- J_y is the total moment of inertia about the yaw pivot
- V_p is the voltage of the pitch motor
- V_y is the voltage of the yaw motor
- θ is the angle about the pitch axis
- ψ is the angle about the yaw axis.

Giving this representation, its parameters are presented in Table 6.2

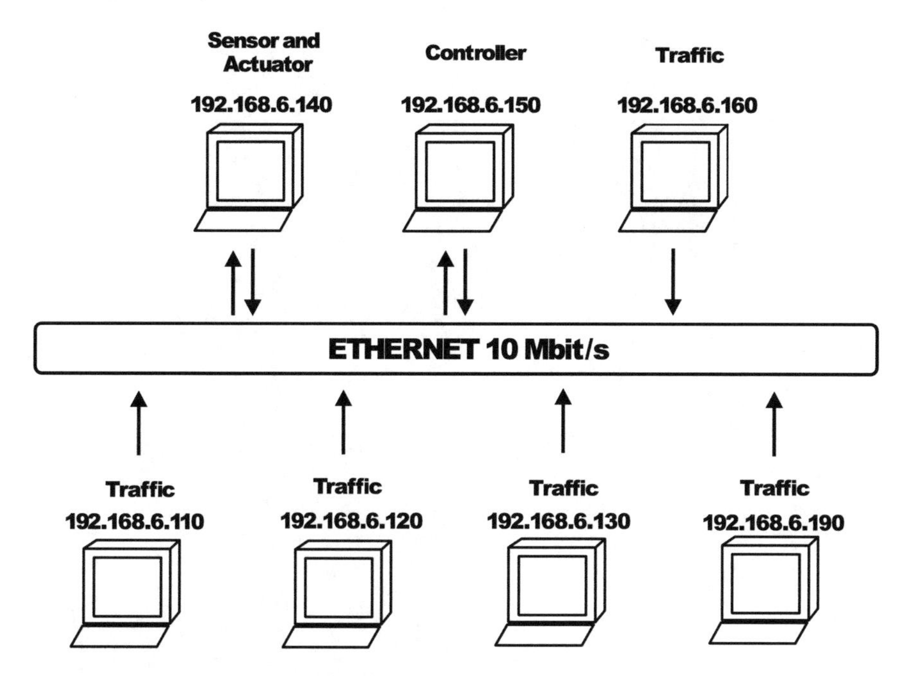

Fig. 6.3 Networked magnetic levitation control system

6.2 Adaptive Fuzzy Control

In this section, the adaptive fuzzy control is designed in a SISO system and MIMO system, the MAGLEV has a configuration with sensor-actuator (S-A) and controller (C) nodes physically distributed and interconnected through a 10 Mb/s Ethernet network with communication protocol UDP/IP (Fig. 6.3).

MAGLEV

The MAGLEV system into the networked control loop is employed to evaluate the performance with the objective of tracking a reference signal [1].

Configuration

The MAGLEV NCS is shown in Fig. 6.4, the computers are Pentium 2 with 254 Mb RAM and an Intel 10/100 Mb Ethernet card. Each computer has an XPC target 2.8v as the operative system, with Matlab 7.1v, and they are connected through a switch Cisco Catalyst 2960, with 24 port 10/100 Mb. The sensors have an A/D Q4 card by Quanser, with 10 bits resolution, and the actuator has a D/A AD512 card by Humusoft, with 8 bits resolution. The sampling period for sensor node is 1 ms, and controller, actuators nodes are event-driven.

Fig. 6.4 Actual implementation of the levitator system within a NCS

The control loop has the following important node types:

- The sensor node is managed with a sampling period T.
- The controller node is event triggered by a sent packet from the sensor node.
- The actuator node is event triggered by a sent packet from the controller node.

The controller design is based on the adaptive fuzzy controller presented in Sect. 4.2 and the fuzzy model explained in Sect. 2.2 where the IF-THEN rules are dependent of the sampling period and the estimated time delays and lost packets (Sect. 2.1).

The control loop has a restriction of precedence messages starting with the sensors, the S-A node sends the sampling variables to the C node and this sends the control signal back to the S-A node to close the loop. Each exchange of information between nodes can be affected by the network performance generating variable time delays or packet losses. The network scheduling is EDF (Earliest Deadline First) without modification. All nodes send UDP (User Datagram Protocol[1]) packets to avoid double traffic in the network. Nevertheless, it is not possible to know whether there is a packet loss or the time delay directly, because there is no acknowledgement that the sent packet has been received, this is a consistent situation along this chapter.

[1]http://www.networksorcery.com/enp/protocol/udp.htm.

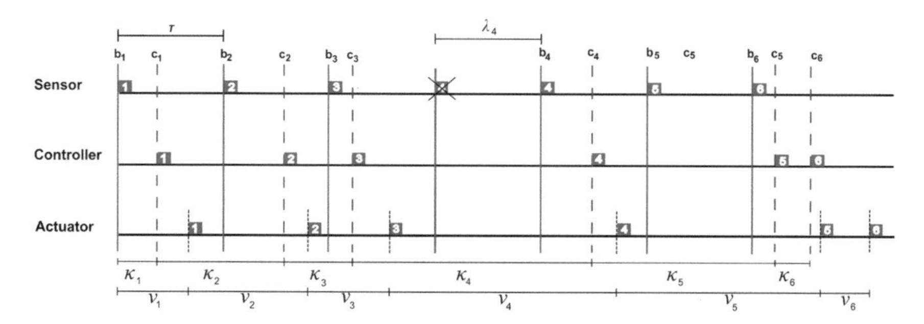

Fig. 6.5 The controller node receives packets from the sensor node, with the information about the states of the system and time stamp of the sent packet

Estimated Variable Time Delay

The estimation of the variable time delay with the configuration S-A and C nodes requires a common timestamp preferably without a clock synchronisation method [1]. This configuration allows estimating the state of the network indirectly since it is possible to establish the time-delays and the lost packets induced by the network using a statistical model.

First, it is necessary to set the measured time and estimated time into the control loop to complete a control feedback is given by the following equation

$$T = t_s + t_{sc} + t_c + t_{ca} + t_a \tag{6.5}$$

where t_s is the time consumed by the sampling, t_{sc} is the communication time between sensors-actuator and controller, t_c is the time it takes to calculate the control signal, t_{ca} is the communication time between controller and sensors-actuator, t_a is the time it takes to the actuator to perform its changes. all the segment times can be estimated or measured except the t_{ca}, affected by the time delays and packet loss in the feedback connection.

So, a statistical model is used to estimate the network-induced time delays and packet loss in the feedback connection, those are seen as a consequence of the variable sampling period $\hat{\tau}_k$ of the NCS from a sensor node to an actuator node. This is estimated by the controller node, to generate a control action. The controller node receives packets from the sensor node, with the information about the states of system $x(k)$ and time stamp of the sent packet b_k. When a packet from the sensor node is received, the time is stamped into controller node c_k, where $k = 1, 2, \ldots$ is the instant when the packet is received. Thus, the sampling period k and packet loss of the controller are calculated using b_k and c_k time stamp (Fig. 6.5).

It is assumed that the internal clocks of the nodes have no drift, or are compensated for, so λ_k is calculated to assume the known controller initial κ_1 sampling period, whereas κ_k is the difference between controller timestamp c_k for $k = 2, 3, \ldots$

$$\lambda_k = round\left(\frac{b_k - b_{k-1}}{T}\right) - 1 \qquad k = 2, 3, \ldots \tag{6.6}$$

$$\kappa_k = (c_k - c_{k-1}) \tag{6.7}$$

While packet loss, λ_k (Eq. 6.6) is the integer number of the difference between the time stamp sensor, divided by the sensor sampling period $T - 1$. Once calculated, λ_k is used to estimate the controller-actuator packet loss $\hat{\lambda}_k^{ca}$, modifying the variable sampling period $\hat{\tau}_k$.

Into the controller the variable sampling period τ_k is calculated with the exponential distribution algorithm as presented in Sect. 2.1:

$$P(\tau) = \begin{cases} \frac{1}{\varphi} e^{-(\tau-\eta)\varphi} & \tau \geq \eta \\ 0 & \tau < \eta \end{cases} \tag{6.8}$$

This function $P[\tau]$ is calculated for each w previous controller sampling period $W = \{\tau_{k-w+1}^{sc}, \ldots, \tau_{k-1}^{sc}, \tau_k^{sc}\}$. First, $q_E = [\eta, \phi]^T$ is obtained by statistical analysis, so the new mean $\eta_{k+1|w}$ is the previous controller sampling period τ_i, with the maximum value of the probability function (Eq. 6.6), and the new standard deviation $\phi_{k+1|w}$ is the root of the variance of the window W.

$$\hat{\eta}_{k+1|w} = w_i | P_{\max_i} [W] \tag{6.9}$$

$$\varphi_{k+1|w} = \sqrt{\text{var}\,(W)} \tag{6.10}$$

So, the variable sampling period $\hat{\tau}_k$ is estimated as:

$$\hat{\tau}_k = \varphi_{k+1|w} + \eta_{k+1|w} + \hat{\lambda}_k^{ca} \quad k = 2, 3..., \tag{6.11}$$

where $\hat{\lambda}_k^{ca}$ is the probability of the controller-actuator packet loss. So, The parameters $[\eta, \ \varphi, \ w]$ are setting using offline measurements of the time delays and packets loss with different traffic loads.

The offline measurements are composed of four parts of 25 ms each. The first part shows the delay with five traffic nodes that are transmitting into the network; the second part has no traffic nodes; the third part has three traffic nodes, and the last part has only one traffic node. So, the average and variance of this data are the initial parameters to construct the statistical model.

$$\begin{aligned} \eta &= 3 \times 10^{-3} \\ \phi &= 1.56 \times 10^{-6} \\ w &= 30 \end{aligned} \tag{6.12}$$

Using these initial parameters a performance test is configurated to estimate the variable sampling period. Figure 6.6 shows the online estimated variable sampling

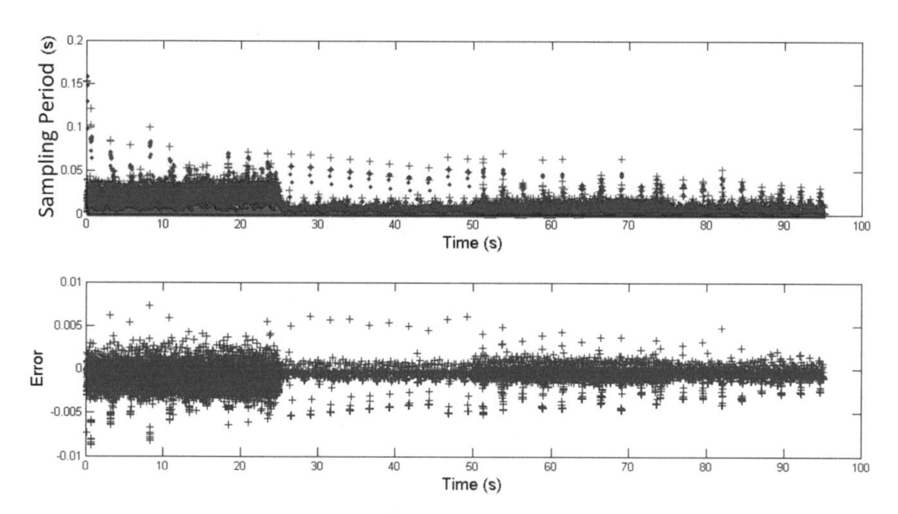

Fig. 6.6 On-line estimated variable sampling period

period (Eq. 6.11), where (+) is the variable sampling period measurement, and $(\cdot)$ is the estimated variable sampling period. The estimation is acceptable with a standard deviation of 1.82×10^{-6}.

Fuzzy Model

The fuzzy model uses the levitation system (Fig. 6.1) with the following nonlinear equations:

$$\begin{aligned} \frac{di(t)}{dt} &= \frac{1}{L} v(t) - \frac{R}{L} i(t) \\ \frac{d^2 a(t)}{dt^2} &= g - \frac{K_e}{m} \frac{i^2(t)}{a^2(t)} \end{aligned} \tag{6.13}$$

where g is the acceleration of gravity, $i(t)$ is the solenoid current, $a(t)$ is the distance between the solenoid to the ball, $v(t)$ is the solenoid voltage, K_e is the magnetic force constant, R is the solenoid resistance, and L is the solenoid inductance.

The nonlinear system of Eq. (6.13) is linearlized with the assumption that the ball is levitating around a nominal operation point (a_0, i_0). The linear model in the state space with the state vector $x(t) = (a(t), \dot{a}(t), i(t))$ is:

$$\dot{x}(t) = \mathrm{A}x(t) + \mathrm{B}u(t) = \begin{bmatrix} 0 & 1 & 0 \\ \frac{K_e i_0^2}{m a_0^3} & 0 & -\frac{K_e i_0^2}{m a_0^2} \\ 0 & 0 & -\frac{R}{L} \end{bmatrix} x(t) + \begin{bmatrix} 0 \\ 0 \\ \frac{1}{L} \end{bmatrix} u(t) \tag{6.14}$$

Each fuzzy rule is generated discretizing the continuous lineal state model (6.14) with a different sampling period using Eqs. (2.33) and (2.34). So, The discrete models are obtained with several sampling periods $\rho = [1,\ 5,\ 10,\ 15]$ ms and four rules have been generated. The first discrete model with $\rho_1 = 5$ ms is:

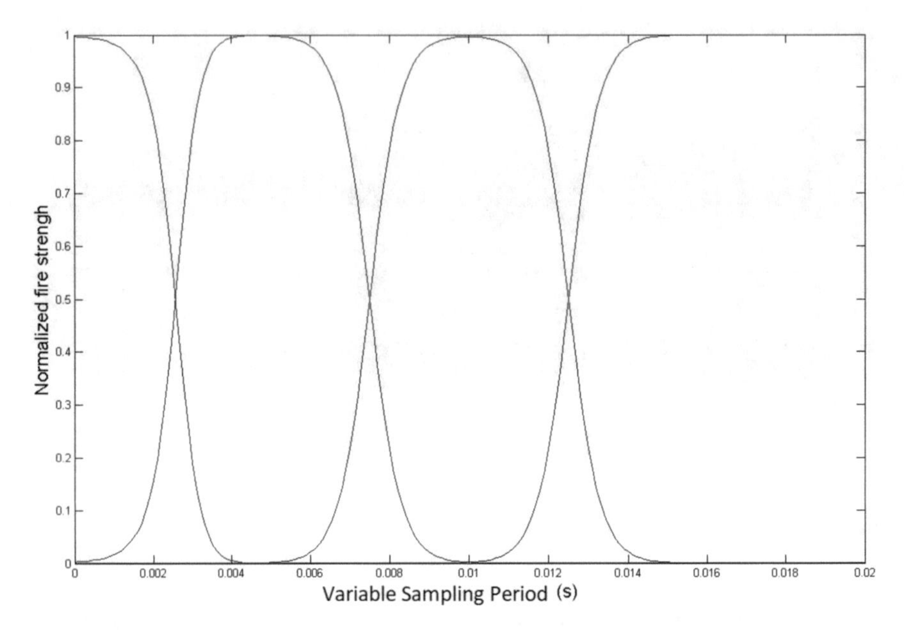

Fig. 6.7 Membership functions for the sampling period with maximum $\overline{h} = 20$ ms

$$x_1(k+1) = \begin{bmatrix} 1.0016 & 0.001 & 0 \\ 3.2718 & 1.0016 & -0.0225 \\ 0 & 0 & -0.9737 \end{bmatrix} x(k) + \begin{bmatrix} 0 \\ 0 \\ 0.0024 \end{bmatrix} u(k) \tag{6.15}$$

The parameters ρ_j, σ_j for $j = 1, \ldots, 4$ membership functions with the bounded $\overline{h} = 20$ ms, are:

$$\begin{aligned} \rho_j &= [0.001 \;\; 0.005 \;\; 0.01 \;\; 0.015] \\ \sigma_j &= [0.0006 \;\; 0.0024 \;\; 0.003 \;\; 0.003] \end{aligned} \tag{6.16}$$

Figure 6.7 shows the membership functions with the parameters designed with a bounded $\overline{h} = 20$ ms. So, the four fuzzy rules are defined as follow:

$$\begin{aligned} &R1: IF \;\; \hat{\tau}(k) \;\; is \; \mu_1 \;\; THEN x_1(k+1) = A_1 x(k) + B_1 u(k) \\ &R2: IF \;\; \hat{\tau}(k) \;\; is \; \mu_2 \;\; THEN x_2(k+1) = A_2 x(k) + B_2 u(k) \\ &R3: IF \;\; \hat{\tau}(k) \;\; is \; \mu_3 \;\; THEN x_3(k+1) = A_3 x(k) + B_3 u(k) \\ &R4: IF \;\; \hat{\tau}(k) \;\; is \; \mu_4 \;\; THEN x_4(k+1) = A_4 x(k) + B_4 u(k) \end{aligned}$$

Fuzzy Control

Along with the fuzzy model, a fuzzy controller is designed following Eq. 4.7 as the control law, so this is designed for each discrete local model. Each feedback control law is carried out by a LQR (Eqs. 4.9 and 4.10). The feedback law for first model is:

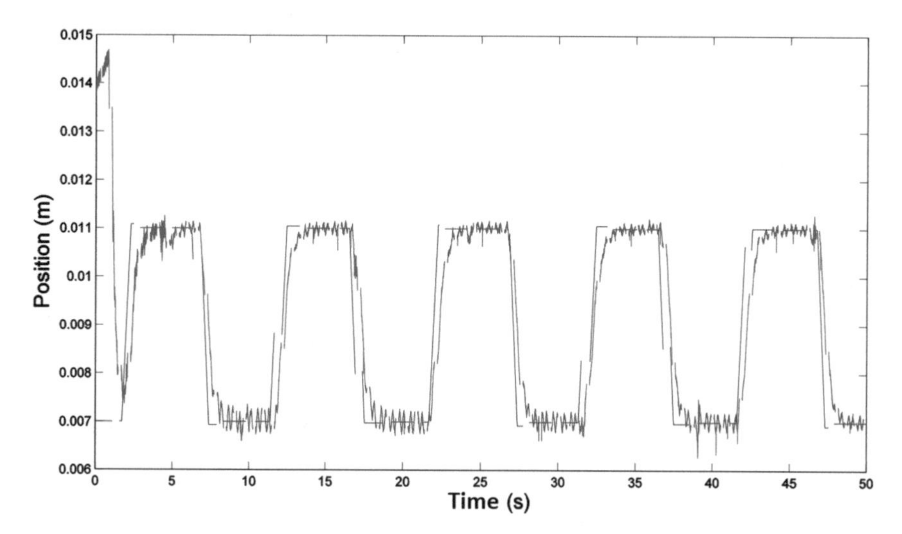

Fig. 6.8 The reference signal is a pulse train

$$K_1 = [-38368 - 103.05411.97] \tag{6.17}$$

The fuzzy controller is implemented with the control loop and five nodes as part of an Ethernet network. The first test is only with the control loop without traffic nodes. The reference signal is a pulse train to show the following trajectories (Fig. 6.8).

The system has an acceptable performance (1.07×10^{-6}) with small fluctuations due to the light sensor, which has no noise filter after the sensor. The controller presents a fine tracking, and a low average steady time of 1.12 s, with an average delay of 3 ms. Although, the variable sampling period (Fig. 6.9) has a range of ($8 \times 10^{-4}, 2 \times 10^{-1}$), exposing a random behaviour similar to the variable sampling period.

To test the efficiency of the fuzzy controller, two case studies are presented: the first case shows stability to delays with three traffic nodes, whereas the second case shows stability to delays and packet loss with five traffic nodes. The message length of traffic nodes is 2 Kbytes, with a sampling period of 1 ms.

The first case (Fig. 6.10) shows the system trajectory with three traffic nodes. The system is stable, with an acceptable performance (1.69×10^{-6}), and small fluctuations even with an average variable sampling period of 6 ms. Notice that there is a maximum variable sampling period of 300 ms, with a 1.98% of packet loss. Therefore, the average settling time is very similar to the case without traffic. This case shows a robust behaviour to moderate time delays and packet loss. However, it is necessary to prove this case with a more aggressive behaviour, this is, with large time delays and more packet loss.

The second case (Fig. 6.11) shows the system trajectory with five traffic nodes that generate large variable sampling periods and a larger packet loss percentage. The system has more fluctuation, but it remains stable with an average settling time of 1.34

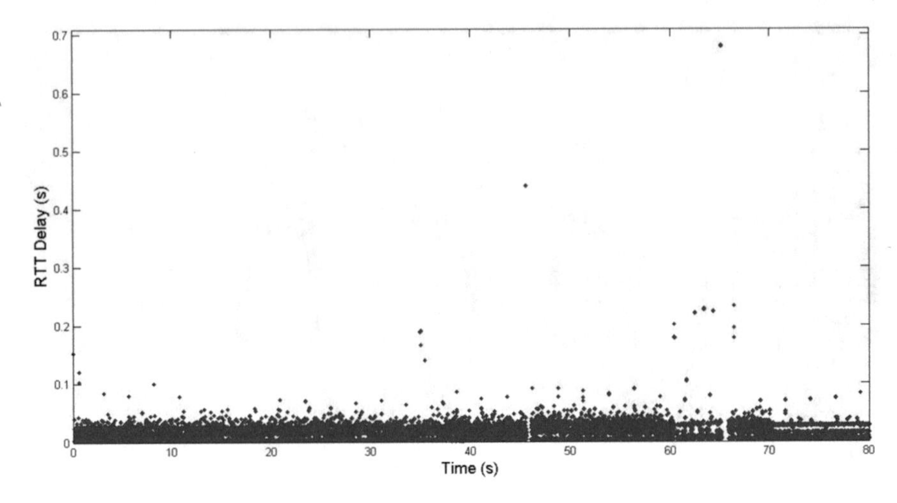

Fig. 6.9 Response to variable sampling period

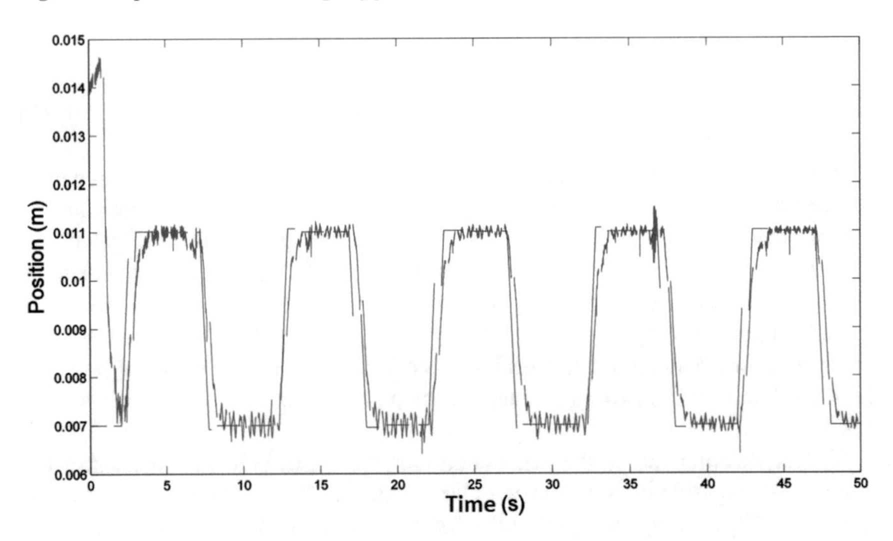

Fig. 6.10 System trajectory based on variable sampling period

ms, similar to the first case. Its performance is fine (2.58×10^{-6}), with an average variable sampling period of 20 ms, and a maximum variable sampling period of 700 ms. The packet loss percentage is 5.7%. The model is obtained from inputs-states data with variable sampling period measurement. The rules are created with an easy clustering method, and the models parameters are updated using backpropagation. It is shown an efficient modelling, with a nonlinear, unstable system.

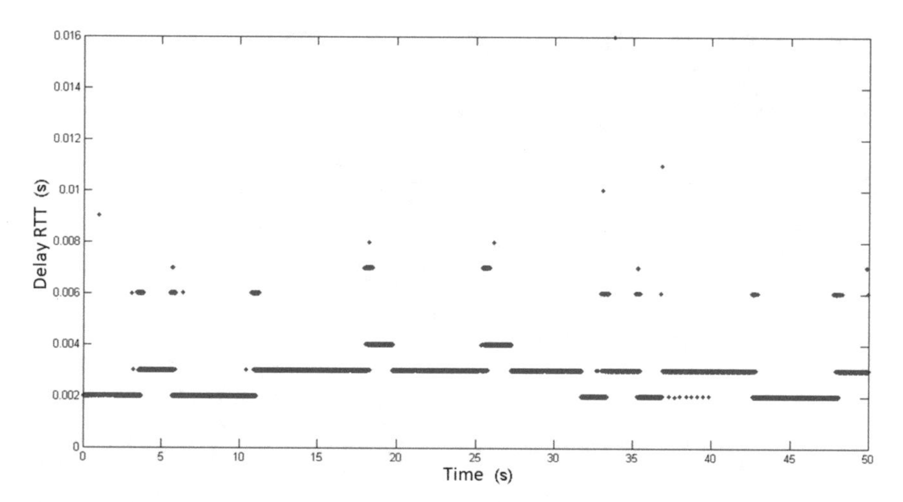

Fig. 6.11 System trajectory with five traffic nodes that generate large varying variable sampling periods

MIMO 2-DOF Helicopter

The 2-DOF helicopter system is integrated to an Ethernet Network (Fig. 6.12) [1]. The computers are PC, those have a Pentium 2 processor with 254Mb RAM and an Intel 10/100Mb Ethernet card; each one has an XPC target 2.8v as an operating system by Matlab 7.1v, and both are connected through a switch Cisco Catalyst 2960 with 24 port 10/100 Mb. The sensors have an A/D Q4 card by Quanser with 10 bits resolution. The actuator has a D/A AD512 card by Humusoft, with 8 bits resolution. The sampling period for the sensor node is 5 ms, and the controller and actuator nodes are event-driven.

The fuzzy control method with variable sampling period is applied to the case study, first, obtaining the behaviour of the variable sampling period, with measurements into an Ethernet network. Once the behaviour sampling period has been established, a probability density function is designed with its mean and standard deviation values as presented in the previous section for levitation experiment. This establishes the range of the measured sampling period, to design the fuzzy model with four fuzzy rules, proposed by the user. Local controllers are designed and used in the fuzzy model to compensate for the varying sampling period. Several tests of network behaviour are used to show the performance of the method. The method is evaluated using a comparison with the performance of the system without a network.

Variable Sampling Period

The first step is estimating the variable sampling period (Eq. 2.4), using an off-line measurement, to obtain the parameters of the probability density function (Eq. 2.1). The off-line measurement is composed of four parts, each one with a duration of 50 s. The first part does not have traffic nodes; the second has one traffic node; the

Fig. 6.12 2-DOF helicopter system integrated to an Ethernet Network

third has three traffic nodes, and the fourth has five traffic nodes. Each traffic node sends packets of 64 bytes, with a sampling period of 1 ms, while the sensor nodes send packets of 64 bytes at 5 ms. With these, the mean and standard deviation of sampling period are obtained, which are the initial parameters of the probability density function.

$$\begin{aligned} \eta &= 6.9 \times 10^{-3} \\ \phi &= 16.3 \times 10^{-3} \\ \omega &= 30 \end{aligned} \tag{6.18}$$

The measurements of the sampling period show that the minimum sampling period is equal to the period of the sensor task (0.5 ms), but although there is no traffic, there is a variation of 20 ms. When there is traffic, the sampling period can reach values in order of hundreds. So, the variation of the sampling period has a very wide range.

Assuming that the periods are independent of each other, it is possible to apply the function of exponential probability density, assuming a low correlation between the sampling periods. With the parameters obtained, the following function is derived:

$$P[\kappa] = \begin{cases} \frac{1}{16.3\times 10^{-3}} e^{-\left(\kappa - 6.9 \times 10^{-3}\right)/16.3 \times 10^{-3}}, & \kappa \geq 6.9 \times 10^{-3} \\ 0, & \kappa < 6.9 \times 10^{-3} \end{cases} \tag{6.19}$$

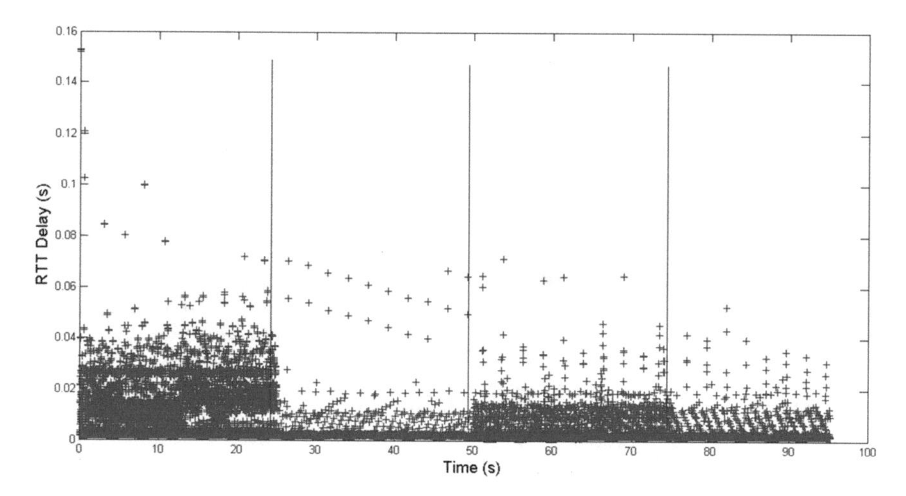

Fig. 6.13 Estimation of sampling period

Off-line data are applied to estimate the sampling period and to justify the use of the exponential function. Figure 6.13 shows the estimation of the sampling period (Eq. 6.11), where $(\cdot)$ is the measured sampling period and $(+)$ is the estimated sampling period. Hence, the estimation is acceptable with an error of 3×10^{-3}. The estimated sampling period is not accurate since the probability depends on the information of the w previous sensor-controller sample periods. However, later, it is shown that it is enough to obtain the best estimate for significant changes in the actuator period, rather than instantaneous changes whose effects can be absorbed. This due to each control signal is generated to be stable over the entire range of the sampling period.

A liability of the selected density function is that it does not estimate the full range of the sampling period because the minimum value of probability density function (6.9 ms) is larger than the period 0.5 ms, established by the task period of the actuators.

Fuzzy Model

Once the range of variation of the sampling period is obtained, the partition of the antecedent part is defined to establish how many fuzzy rules have to be generated. This provides the parameters for membership functions. The consequent part is obtained by the continuous linear state space model of the 2-DOF helicopter linearizing the non-linear equations in Eq. 6.4 about the quiescent point ($\theta_0 = 0$, $\psi_0 = 0$, $\dot{\theta}_0 = 0$, $\dot{\psi}_0 = 0$), and with states $x = [\theta \ \psi \ \dot{\theta} \ \dot{\psi}]^T$, and inputs $u = [V_p \ V_y]^T$. The state space model is:

$$\dot{x} = \begin{bmatrix} 0\,0 & 1 & 0 \\ 0\,0 & 0 & 1 \\ 0\,0 & -\frac{B_p}{J_p+ml^2} & 0 \\ 0\,0 & 0 & -\frac{B_y}{J_y+ml^2} \end{bmatrix} x(t) + \begin{bmatrix} 0 & 0 \\ 0 & 0 \\ \frac{k_{pp}}{J_p+ml^2} & \frac{k_{py}}{J_p+ml^2} \\ \frac{k_{yp}}{J_p+ml^2} & \frac{k_{yy}}{J_p+ml^2} \end{bmatrix} u(t)$$
$$y = \begin{bmatrix} 1\,0\,0\,0 \\ 0\,1\,0\,0 \end{bmatrix} x \tag{6.20}$$

For each fuzzy rule, the continuous linear model is discretized with a sampling period $T_j, j = 1, \ldots, r$, defined for the case study as $T = [0.005\ 0.010\ 0.015\ 0.020]^T$. The discrete local model for $T_1 = 0.005$ used for the first rule is:

$$x_1(k + 0.005) = \begin{bmatrix} 1\ 0\ 0.0048 & 0 \\ 0\ 1 \quad 0 & 0.0049 \\ 0\ 0\ 0.9547 & 0 \\ 0\ 0 \quad 0 & 0.9827 \end{bmatrix} x(k) + \begin{bmatrix} 2.913\text{E}-5 & 9.722\text{E}-7 \\ 2.995\text{E}-6 & 9.834\text{E}-6 \\ 1.156\text{E}-2 & 3.859\text{E}-4 \\ 1.195\text{E}-3 & 3.922\text{E}-3 \end{bmatrix} u(k) \tag{6.21}$$

The antecedent parameters and the maximum bounds are defined by the user, using information from the off-line sampling period measurement. So, the parameters ρ_j and σ_j for $j = 1, \ldots, r$ with the maximum bounded $\nu^{MAX} = 25$ ms, are:

$$\begin{aligned} \rho_j &= \begin{bmatrix} 0.005\ 0.01\ 0.015\ 0.02 \end{bmatrix} \\ \sigma_j &= \begin{bmatrix} 2\times10^{-3}\ 2\times10^{-3}\ 2\times10^{-3}\ 2\times10^{-3} \end{bmatrix} \end{aligned} \tag{6.22}$$

Notice that there are three ($\mu = 3$) regions S_s with two overlapped membership functions (Fig. 6.14). Therefore, there should be only three matrices P_s, with three conditions each, to ensure stability, contrary to the general method that requires 16 conditions for a common matrix P.

Fuzzy Control

With the set of fuzzy rules for the fuzzy model, an LQR fuzzy controller (Eq. 2.31) is designed for each local model. Two integrators are added as states, to reduce the steady-state error. All feedback laws are designed by LQR (Eq. 4.7). The feedback law for the first local model is:

$$K_1 = \begin{bmatrix} 15.4 & 1.53 & 4.91 & 0.677 & 12.2 & 0.718 \\ -1.97 & 17.3 & -0.241 & 6.2 & -1.24 & 7.03 \end{bmatrix} \tag{6.23}$$

Once the controller is designed, experiments are conducted to test the performance and robustness to time delays and packet loss. There are three tests: the first test shows the performance of the NCS with a normal behaviour without external traffic on the

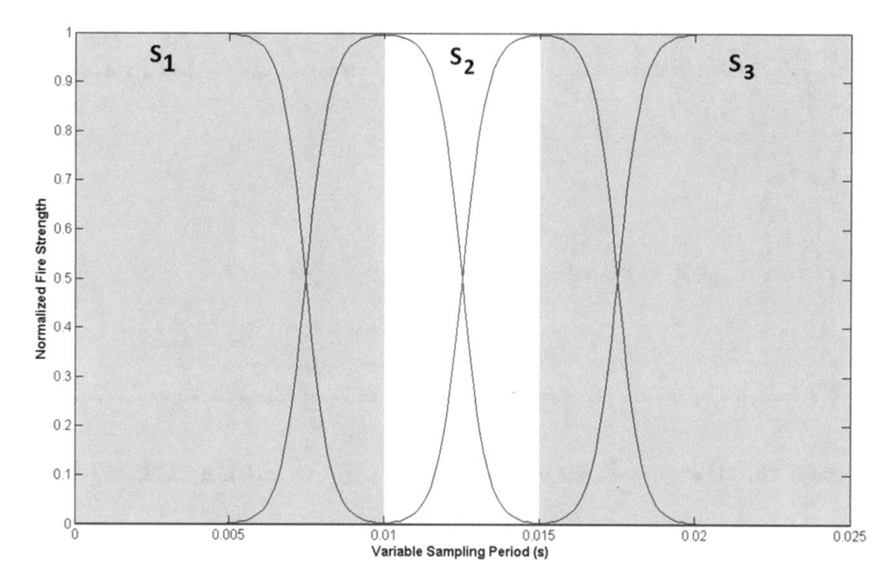

Fig. 6.14 Three regions S_s with two overlapped membership functions

network. The second test introduces four nodes to generate traffic on the network and shows the control performance under variable sampling periods and some amount of data loss. Finally, a third test presents the performance of the system to six traffic nodes generating longer variable sampling periods and a greater amount of packet loss.

The first test is carried out only with the control loop without network traffic. The reference signal is the zero position of pitch angle (θ), and an initial position of $-41°$ (Fig. 6.15). Applying the controller, experiments are performed, in which the response is always stable with an average steady-state error of $0.2°$, and an overshoot of $1°$. Moreover, the behaviour of fuzzy control is compared with the digital control without the network obtaining similar behaviour.

In this test, the behaviour of the variable sampling period and the packet loss are shown in Fig. 6.16, where the range of $\hat{\upsilon}$ is (0.0005, 0.021), with a mean 7 ms, and 0.03% of packet loss. The integral absolute error (IAE) is 1.1762. This is similar to the performance of the digital control, that is 1.0652.

The second test shows a robust behaviour of the system with four traffic nodes into an Ethernet network. The traffic packets have 254 bytes of length, and 1 ms as sampling period.

Figure 6.17 shows the behaviour of the system during the second test. The pitch position with fuzzy control (θ) presents a small oscillation during the transition, which vanishes as time passes, whereas the position with feedback control presents more oscillations and a steady state error of $3°$. All experiments with fuzzy control are stable, with an average steady-state error of $0.5°$. Although, the variable sampling period has a maximum of 300 ms, and an average of 9 ms, with a data loss of 0.34%.

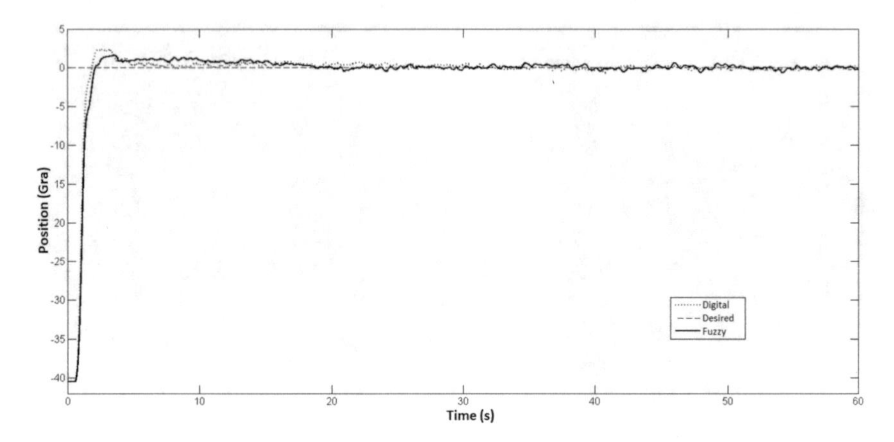

Fig. 6.15 The reference signal: zero position of pitch angle (θ), and an initial position of $-41°$

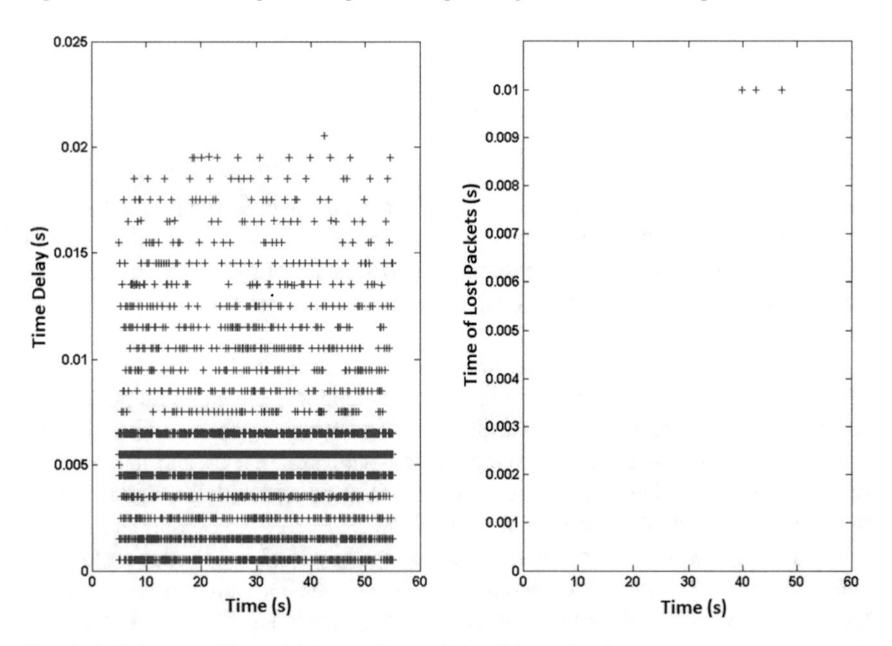

Fig. 6.16 Behavior of the variable sampling period and the packet loss

The maximum IAE of the fuzzy control is 1.1332, better than the feedback control performance of 1.8043.

The third test shows the system behaviour with six traffic nodes sending packets into the network (Fig. 6.18). The response with fuzzy control presents more oscillation during the transition state, which vanishes as time passes. The feedback control performance presents larger oscillations, with bad performance. All experiments with the fuzzy control are stable, with a maximum overshoot of 2.5°, and an aver-

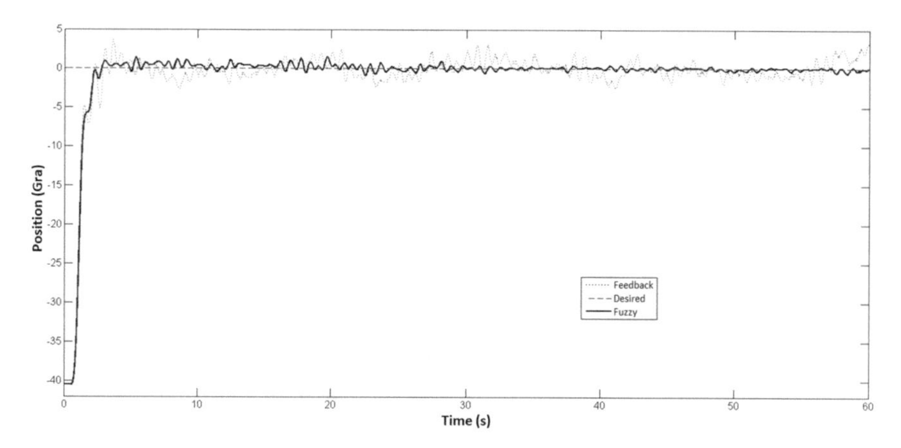

Fig. 6.17 Response with 0.34% packet loss and maximum variable sampling period of 300 ms

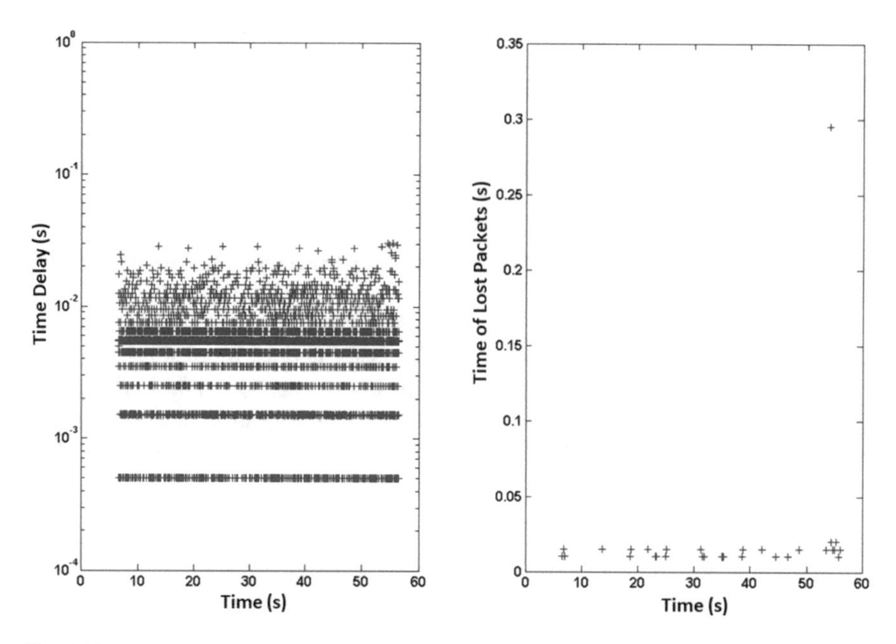

Fig. 6.18 System with six traffic nodes sending packets into the network

age steady-state error of 0.4°. The fuzzy controller compensates for a maximum sampling period of 200 ms, and 1.84% packet loss (Fig. 6.19). The maximum performance of the fuzzy control is 1.3463, better than the performance of the feedback control (3.889), due to the later does not compensate for time delays and packet loss.

The NCS performance with fuzzy control is almost the same as the performance of the digital control, without external network traffic. Further, it is more robust to sampling period variations and packet loss. This approach maintains the system

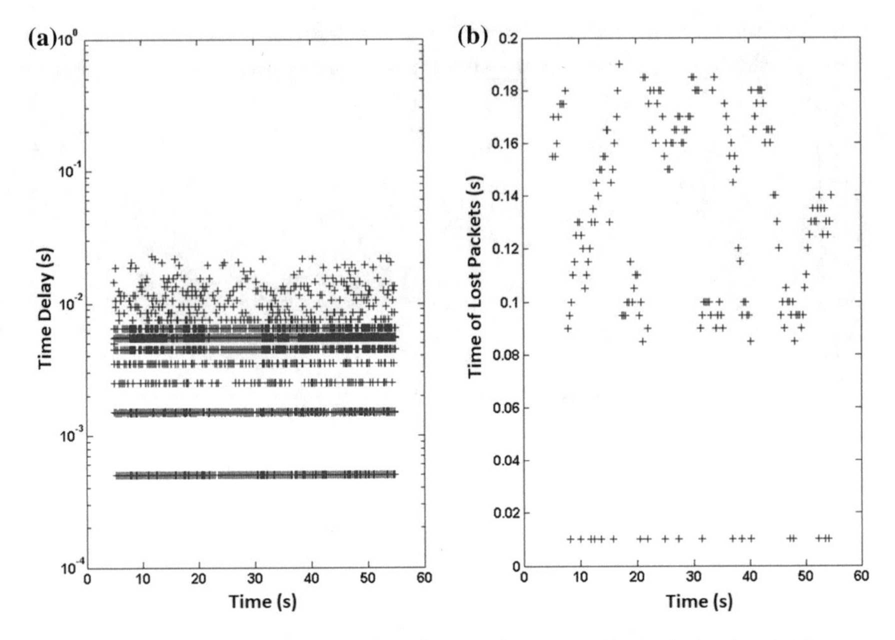

Fig. 6.19 The fuzzy controller compensates for a maximum sampling period and packet loss

stable, with a good performance to a wide range of variations and a considerable packet loss.

Figure 6.20 compares the performance of the fuzzy control and feedback control through time, observing the behaviour of the fuzzy control in all three tests (no traffic, four, and six traffic nodes), and taking the lower response for the performance of digital control. Notice that the performance of the fuzzy control on the three tests remains always close to the digital control performance, even with variable time delays and packet losses.

6.3 Sampling Frequency Application

In order to prove the effectiveness of the method proposed in Sect. 4.3, the SISO MAGLEV and the MIMO 2DOF Helicopter case studies are used, both connected to the same network. Thus, both share the network as part of their control loops. Each case study is presented as a state variable model, to generate the fuzzy models and the feedback matrices for the fuzzy control. The important issue of this method is the frequency transition strategy. This strategy modifies the transmission periods of the sensors in both systems: four sensors for the 2DOF Helicopter, and two sensors for the MAGLEV.

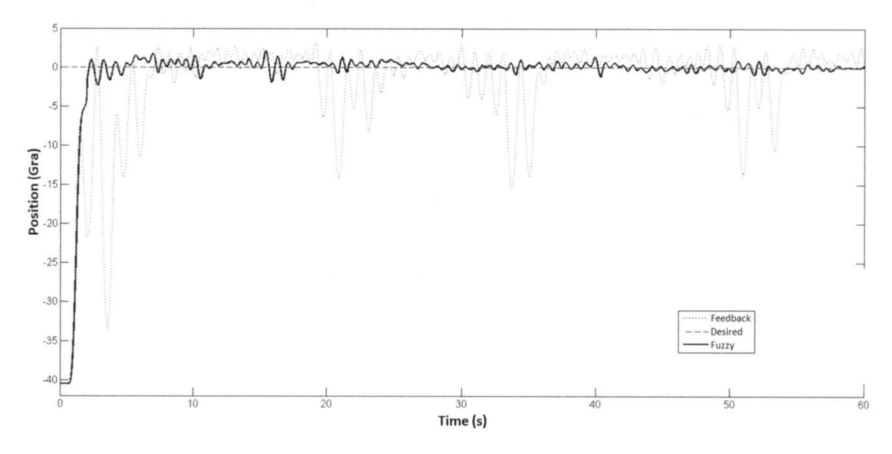

Fig. 6.20 Comparison of the performance of the fuzzy control and feedback control through time, in all three tests

MIMO Helicopter

Fuzzy Model

Using the state-space $x = (\theta, \psi, \dot{\theta}, \dot{\psi})$, three discrete models are obtained by the continuous linear state space model of the 2-DOF helicopter, linearizing the non-linear equations in Eq. 6.4, first using the quiescent point $(\theta_0 = 0, \psi_0 = 0, \dot{\theta}_0 = 0, \dot{\psi}_0 = 0)$, and with states $x = [\theta\ \psi\ \dot{\theta}\ \dot{\psi}]^T$, and inputs $u = [V_p\ V_y]^T$. The state space model is:

$$A_1 = \begin{bmatrix} 0 & 0 & 1 & 0 \\ 0 & 0 & 0 & 1 \\ 0 & 0 & -9.2593 & 0 \\ 0 & 0 & 0 & -3.4868 \end{bmatrix} \tag{6.24}$$

$$B_1 = \begin{bmatrix} 0 & 0 \\ 0 & 0 \\ 2.3611 & 0.0787 \\ 0.2401 & 0.7895 \end{bmatrix} \tag{6.25}$$

$$C = \begin{bmatrix} 1 & 0 & 0 & 0 \\ 0 & 1 & 0 & 0 \end{bmatrix} \tag{6.26}$$

Now linearizing about $x = (40, 0, 0, 0)^T$:

$$A_2 = \begin{bmatrix} 0 & 0 & 1 & 0 \\ 0 & 0 & 0 & 1 \\ 18.8362 & 0 & -9.2768 & 0 \\ 2.5148 & 0 & 0 & -4.4618 \end{bmatrix} \tag{6.27}$$

$$B_2 = \begin{bmatrix} 0 & 0 \\ 0 & 0 \\ 2.3667 & 0.0789 \\ 0.3073 & 1.0102 \end{bmatrix} \quad (6.28)$$

$$C = \begin{bmatrix} 1\,0\,0\,0 \\ 0\,1\,0\,0 \end{bmatrix} \quad (6.29)$$

and linearizing about $x = (-40,\ 0,\ 0,\ 0)^T$:

$$A_3 = \begin{bmatrix} 0 & 0 & 1 & 0 \\ 0 & 0 & 0 & 1 \\ -18.8362 & 0 & -9.2768 & 0 \\ -2.5148 & 0 & 0 & -4.4618 \end{bmatrix} \quad (6.30)$$

$$B_3 = \begin{bmatrix} 0 & 0 \\ 0 & 0 \\ 2.3667 & 0.0789 \\ 0.3073 & 1.0102 \end{bmatrix} \quad (6.31)$$

$$C = \begin{bmatrix} 1\,0\,0\,0 \\ 0\,1\,0\,0 \end{bmatrix} \quad (6.32)$$

The three discrete models are designed for covering the range of minimum, maximum, and middle pitch state. Now, in terms of fuzzy control, three rules are proposed around these three quiescent points, where control gains are provided by Eqs. 6.33, 6.34 and 6.35

Fuzzy Control

The design of the fuzzy controller is carried out using Eq. 3.38, which represents a LQR+I control for each fuzzy rule. In each fuzzy rule, the antecedent part is represented by pitch state value, around the nominal value of the respective linearization:

- Rule 1:
  IF x_1 is about 0º,
  THEN $u = -F_1 x$
- Rule 2:
  IF x_1 is about 40º,
  THEN $u = -F_2 x$
- Rule 3:
  IF x_1 is about −40º,
  THEN $u = -F_3 x$

From these fuzzy rules, the following feedback matrices are constructed:

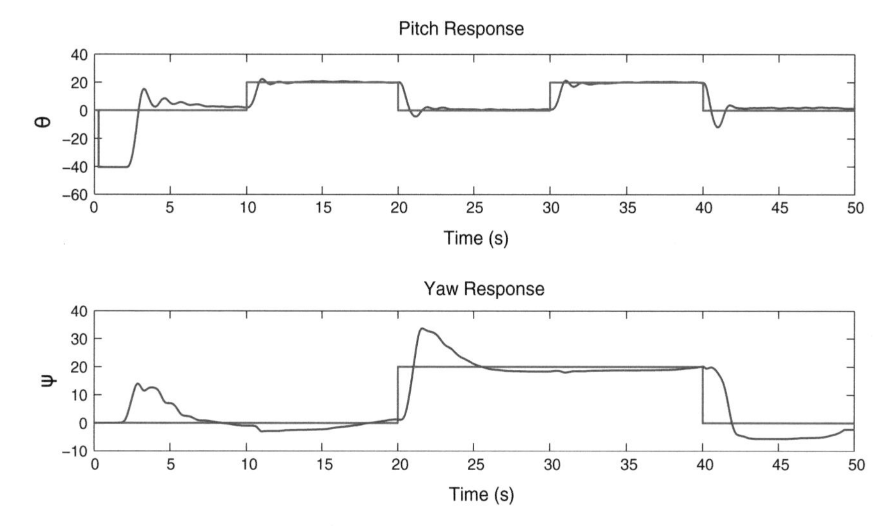

Fig. 6.21 Pitch and Yaw angle when frequencies are whitin the bounds

$$F_1 = \begin{bmatrix} 18.9 & 1.98 & 7.48 & 1.53 & 7.03 & 0.770 \\ -2.22 & 19.4 & -0.45 & 11.9 & -0.770 & 7.03 \end{bmatrix} \tag{6.33}$$

$$F_2 = \begin{bmatrix} 66.2043 & 6.5036 & 9.4245 & 1.5276 & 67.1087 & 7.3142 \\ -8.2791 & 71.8961 & -1.0251 & 15.3362 & -8.8907 & 73.2738 \end{bmatrix} \tag{6.34}$$

$$F_3 = \begin{bmatrix} 147.3617 & 20.0491 & 30.0521 & 4.6301 & 180.7394 & 22.1538 \\ -17.5406 & 203.6146 & -2.9301 & 45.5026 & -19.7219 & 217.6437 \end{bmatrix} \tag{6.35}$$

Once the fuzzy controller has been designed, the fault parameters are added, represented by Eqs. 3.40 and 3.41. From these, the frequency transition strategy is designed, along with Eqs. 3.18 and 3.19. The experiments take into consideration the frequency transition as a reconfiguration strategy to minimise the effects of time delays and packet loss.

The first experiment shows the performance of the system in normal operation, this is, without network traffic. The other experiments consider different levels of network traffic, which impact on the system with different levels time delays and packet loss.

The results of the first experiment are presented in Fig. 6.21. The twin-rotor presents a suitable response, even with frequency transitions. The red colour signal represents the desired pitch and yaw angle, whereas the blue colour signal is the obtained response.

Notice that the transmission frequencies are outside the lower bound of the schedulability region during short instants.

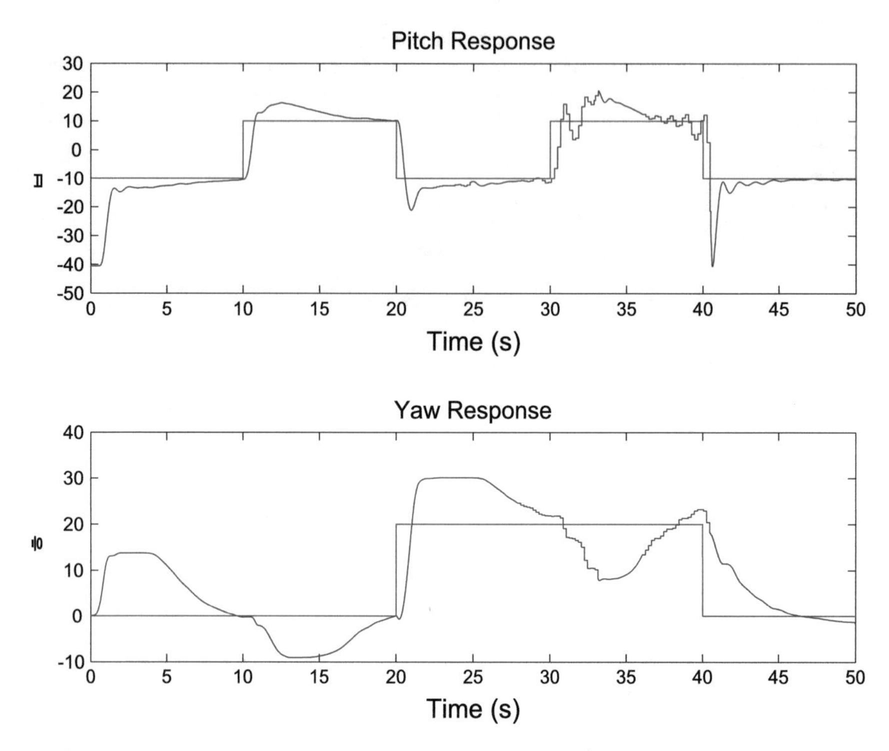

Fig. 6.22 Pitch and Yaw angle when frequencies by instants are not within the bounds

For the second experiment, in which a level of network traffic is introduced during the whole experiment, Fig. 6.22 shows the following results: at $t = 10$ and $t = 30$, the frequencies leave the schedulability region for 5 s, and then return to the region. This is due to the reconfiguration strategy, which implies a frequency transition.

The control design and the frequency transition design are similar to the MIMO 2-DOF Helicopter case study, presenting three fuzzy rules using Eq. 3.38, and five frequency transition rules, presented in Eq. 3.19. Regarding only the helicopter system response, this is presented in Fig. 6.23, where frequency transition is shown in Fig. 6.24 with the related frequency variations on sensors from the helicopter and magnetic levitation systems. Once each control loop has been designed, the performance of each NCS is analysed, along with the frequency transition strategy. Notice that this performance is stable in all experiments.

Now, in terms of magnetic levitation system, its response is presented in Fig. 6.25, where clearly there is no variation during time delays.

Observe that the magnetic levitation system response, in terms of the current value and ball position during the frequency transmission change, is a pulse train as presented in Fig. 6.26, where the red colour signal is the desired current and ball position, and the blue colour signal is the system response.

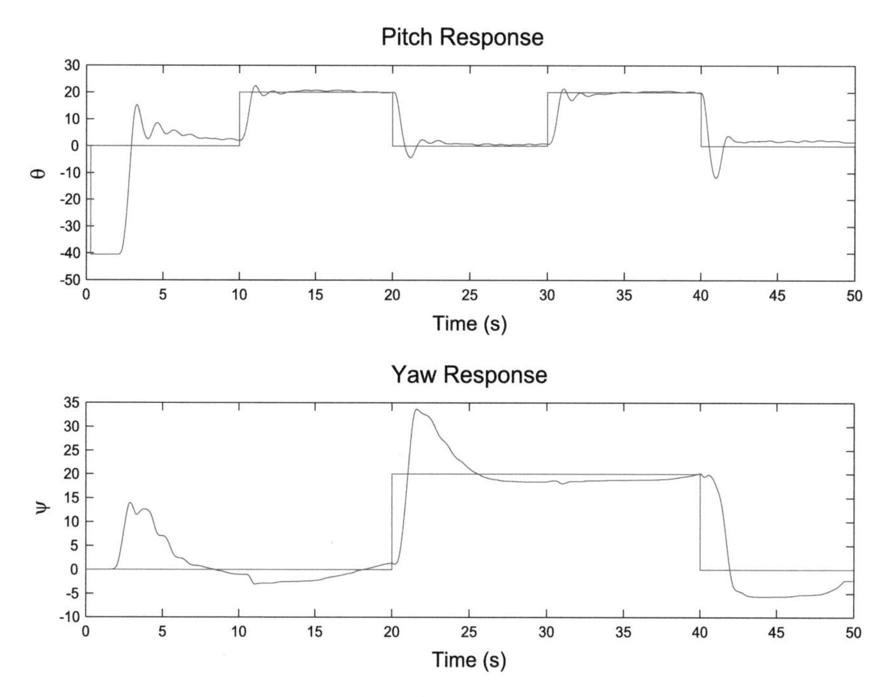

Fig. 6.23 Helicopter response in frequency transitions

The next result is obtained (Fig. 6.27) when the frequencies change at $t = 25$ and $t = 40$, slightly leaving the schedulability region for 3 s and 5 s, respectively. Then, the frequencies return within the region. This experiment allows observing that the system performance against changes in the set points, and allows to analyse the performance of the frequency transition algorithm, which returns frequencies into a bounded region just after a perturbation generated by the network traffic and/or bandwidth decrease.

6.4 2-DOF Helicopter System Codesign

In previous experiments, information about the system's performance is used in the fuzzy controller compensates network generated uncertainties. However, another way to compensate network uncertainties relays on decreasing the network traffic. This requires to carry out some adjustments, such as scheduling packet emission, increasing communication speed, diminishing the transmission rate between nodes, decreasing packet size, and so on. Of all these, the simplest to locally implement and design in each of the nodes is to diminish the transmission rate between nodes. A smaller packet transmission for each time unit tends to decrease the traffic while providing more time for the packet to arrive at its destination.

The fuzzy scheduler is shown in Sect. 4.4 aims to modify the sampling periods during transmission of the NCS, depending on the system and network performance.

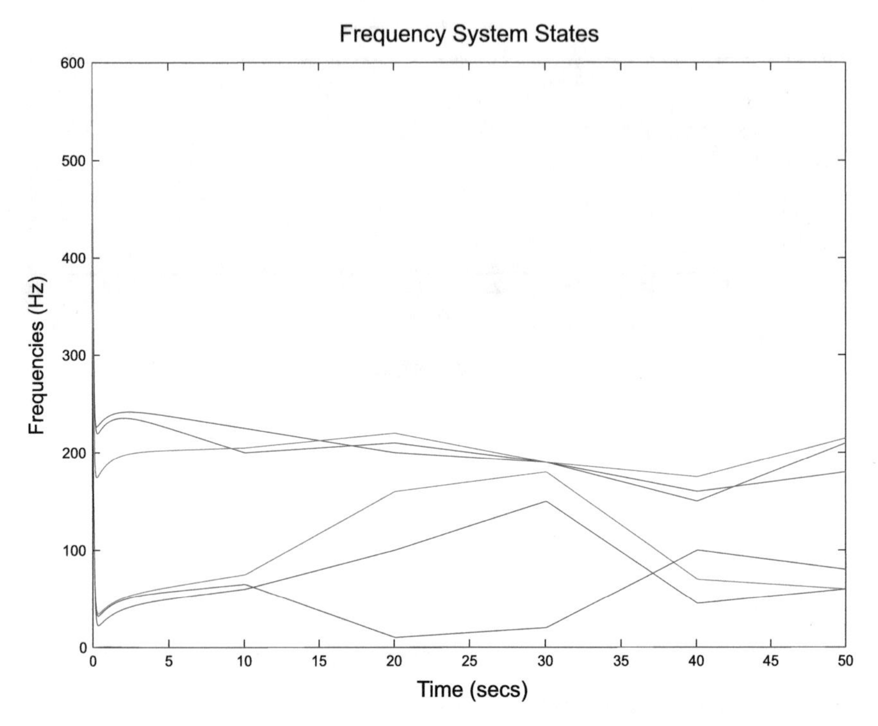

Fig. 6.24 Frequency transition. The red lines represent the 2-DOF Helicopter, while the blue lines represent the magnetic levitation system

The fuzzy scheduler design completes the controller-scheduler codesign, in which the controller has been designed for a 2-DOF Helicopter simulator, described in Sects. 3.5 and 6.2. For designing and validating the proposed fuzzy scheduler, the same simulator with the fuzzy controller (designed in Sect. 6.2) is used, as well as a completely distributed configuration.

The general objective of codesign is, first, controlling the pitch and yaw positions against variations of network conditions, and second, scheduling the transmission within the NCS. The particular objective of the fuzzy controller is to generate an adequate control signal to compensate the variant uncertainties of the network, while the particular objective of the fuzzy scheduler is to modify the transmission period of the NCS, aiming to improve the QoS, and with no degradation of the QoC.

The fuzzy scheduler design shown in Sect. 4.4 is divided into two stages: a first stage that implements a module in the actuator node, measuring both QoS and QoC for each scheduling period, and a second stage that implements a module in the sensor node with the fuzzy scheduler, calculating the next sampling period of the NCS, according to QoS and QoC. Particularly, QoC is measured based on the MAE, and QoS is measured based on the number of lost deadlines.

Depending on the performance measures of the system and the network, the rank of MAE for the scheduler design is $[0,\ 1.4]$, while regarding the percentage of lost

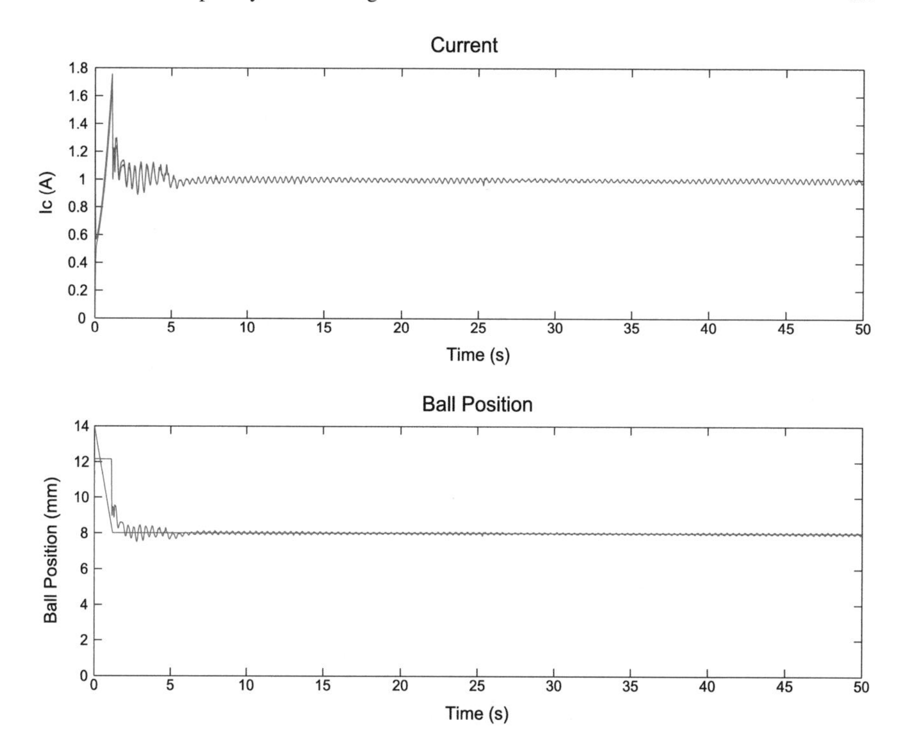

Fig. 6.25 Levitator response in frequency transitions

deadlines, the rank is [0, 1], with a maximum deadline of $h_{max} = 30$ ms. All this with a scheduling period of $\delta = 1$ s. For design purposes, four fuzzy memberships are chosen within each rank, with an output rank for the new sampling period of [0.005, 0.01, 0.015, 0.02]. With these, 16 fuzzy rules are heuristically generated from a knowledge database.

Experiments have been designed to show the applicability of codesign, aiming to follow a square signal for the pitch position, and observing the behaviour of uncertainties and the system's performance. Each experiment involves the generation of traffic at different levels between elapsed times of 50 and 80 s. For comparative purposes, a hybrid controller [2] is designed for generating a feedback control law that assures stability against network uncertainties within the application rank.

The first experiment shows the performance of the codesign approach, with normal operation conditions, and without traffic on the network. In this case, the performance of both, codesign approach and hybrid controller approach, fulfil the control criteria. The pitch position for the fuzzy codesign approach (Fig. 6.28a) keeps an error during average stable state $e_{ss} \approx 0.5$, with a steady time $T_s = 3$ s, and an overshot $\zeta = 0$. The response of the hybrid controller approach is very similar, and it is not shown in the figure.

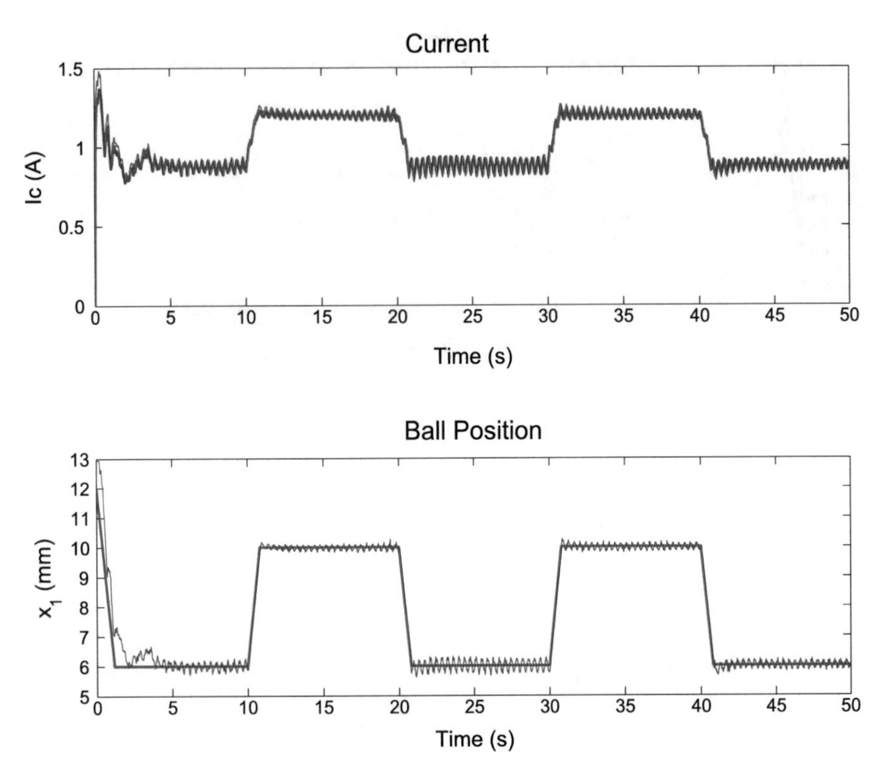

Fig. 6.26 Current and ball position when frequencies are within the bounds

Figure 6.28b shows the control signal of the pitch, which keeps a behaviour within the normal operation rank with variations due to the uncertainties of the network. These variations are compensated by the fuzzy controller.

Regarding the behaviour of the network without external traffic (Fig. 6.29), there are only time delays with a maximum of 4 ms, and an average of 1 ms. Nevertheless, the most important issue is that there are no lost deadlines.

Observing the system's and network's behaviour, the fuzzy scheduler, as the complement of the NCS codesign, evaluates its performance at each scheduling period $\delta = 1$ s, and defines the next sampling period h. Figure 6.30a shows the system's performance measured through the MAE, while the performance of the network is measured using a percentage of lost deadlines (Fig. 6.30b), both during each scheduling period. It also shows the behaviour of the fuzzy scheduler, that obtains the next sampling period in a rank of [0.005 0.02] s, decreasing when there are large errors, and increasing when the error is small and requires just a few control changes by time unit.

Aiming to observe the effects of NCS codesign when there are variations due to network uncertainties, traffic is generated at different levels, and thus, measuring the

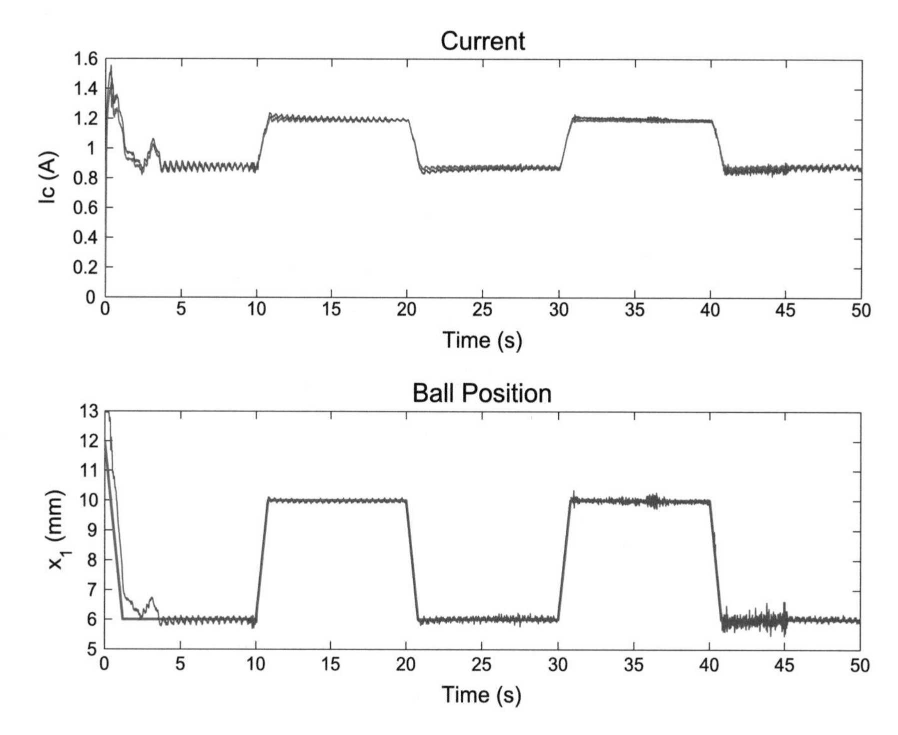

Fig. 6.27 Current and ball position when frequencies by instants are not within the bounds

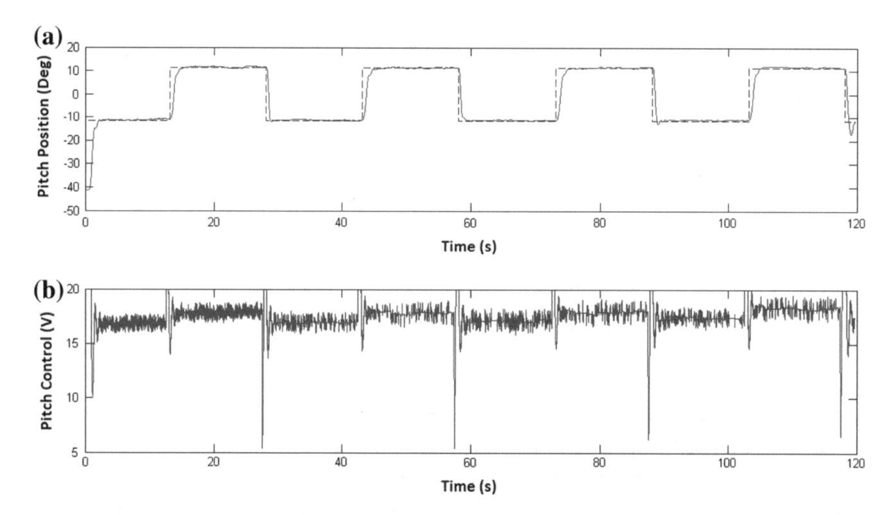

Fig. 6.28 **a** Performance of the codesign approach and **b** control signal, without external network traffic

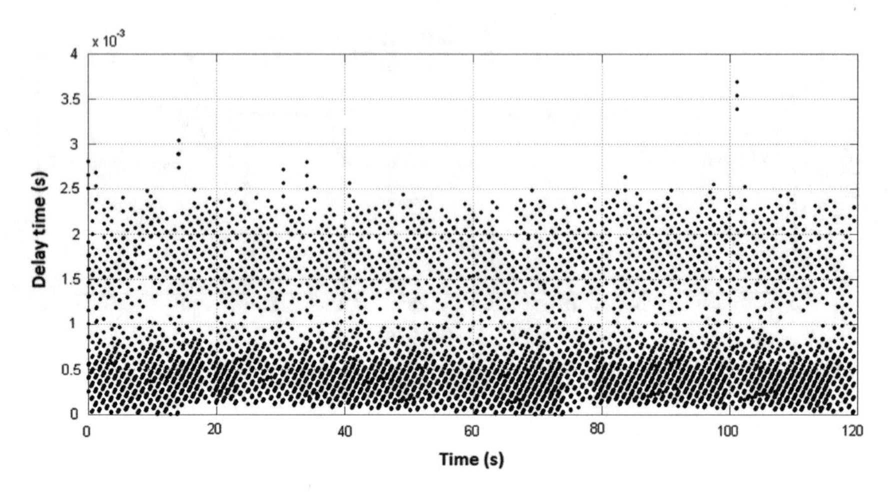

Fig. 6.29 Behavior of the network without external traffic for the NCS codesign

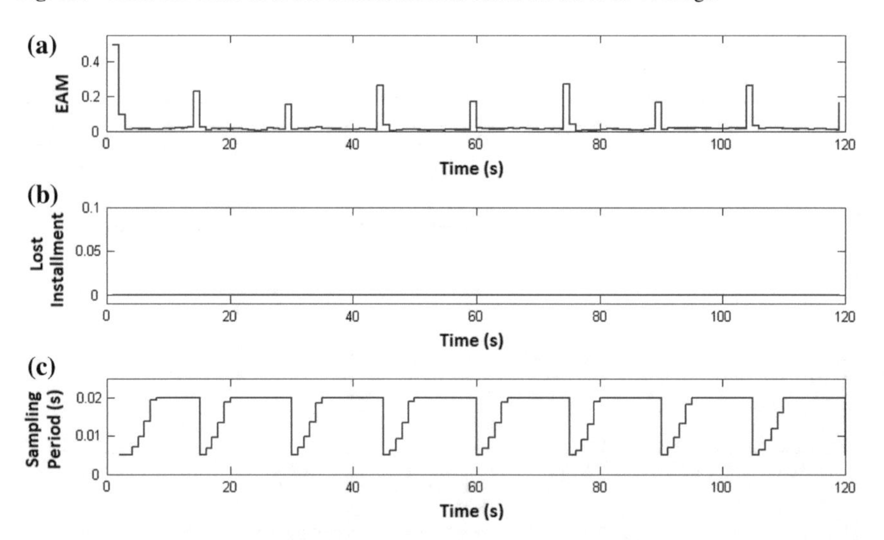

Fig. 6.30 **a** MAE, **b** lost deadlines, and **c** new sampling period for the NCS, without external traffic

system's behaviour with the fuzzy controller, as well as the network behaviour and the sampling period, applying the fuzzy scheduler.

First, some light traffic is applied within the rank of [50 80] s. The performance of both, the fuzzy codesign and the hybrid controller is kept stable. Codesign (Fig. 6.31a) exhibits an overshot of $\zeta = 4°$, with an average error at stable state $e_{ss} \approx 0.5$, and a steady time $T_s = 3$ s. However, the degrading effects between times 50 s and 80 s are minimal. On the other hand, the hybrid control (Fig. 6.31b) degrades its performance when there are network traffic variations, with an overshoot of $\zeta = 34°$, an average error at stable state $e_{ss} \approx 2$, and a steady time $T_s = 10$ s.

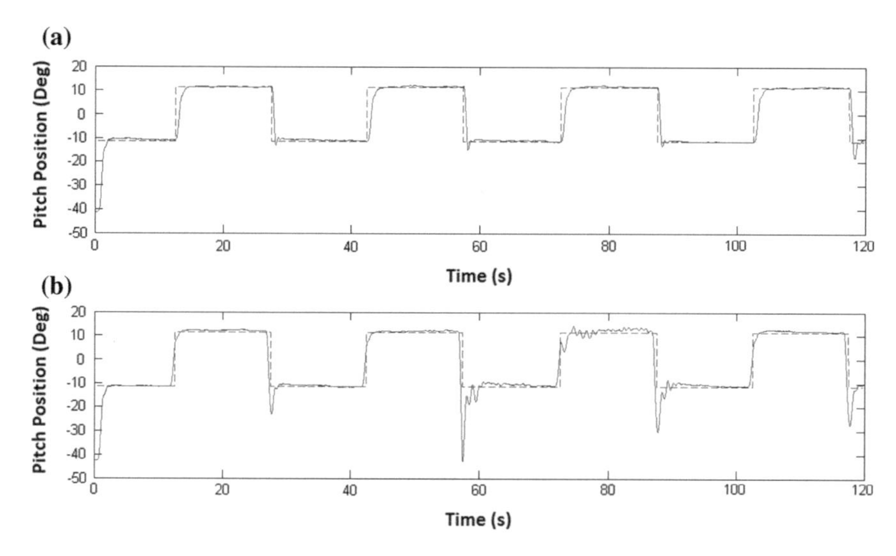

Fig. 6.31 Performance of **a** the fuzzy codesing and **b** hybrid control, with light traffic (50–80 s)

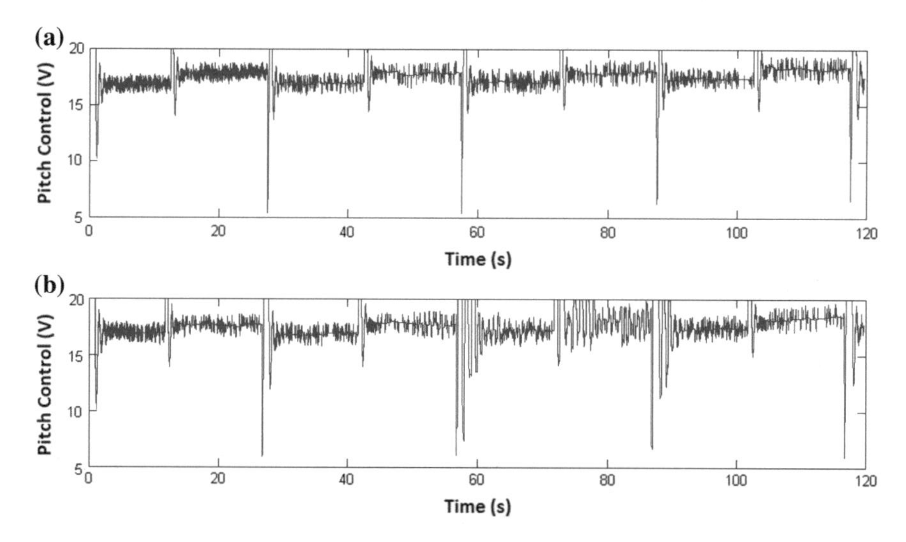

Fig. 6.32 Control signal of **a** the fuzzy codesign and **b** the hybrid control, with light traffic

The behaviour of the control signals for pitch angle are shown in Fig. 6.32. The traffic changes at 50 s. Both approaches no reflect the change of traffic in their performance neither in the control signal, until there is a reference change that points an error increment at 58 s. The difference is that the fuzzy codesign (Fig. 6.32a) performs only little changes in the control signal, while the hybrid control (Fig. 6.32b) carries out abrupt changes due to the lack of information because the lost deadlines that are not taken into account when calculating the control signal.

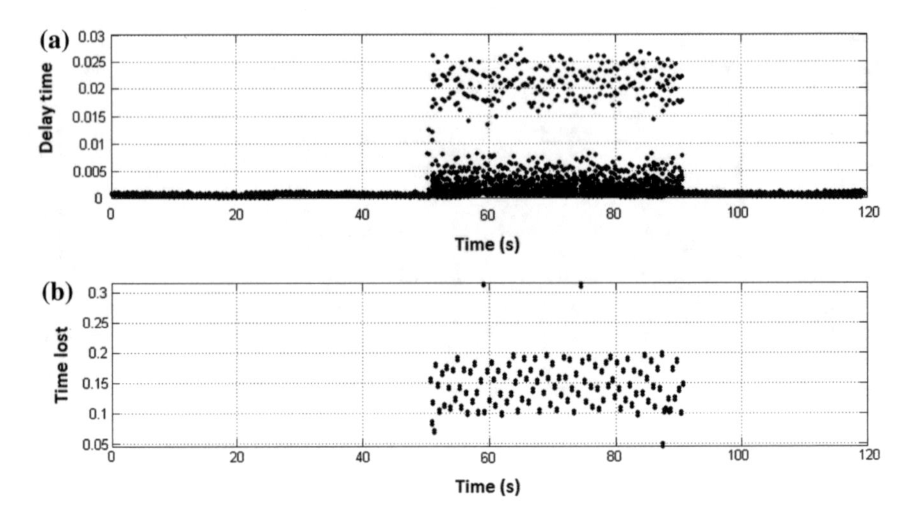

Fig. 6.33 **a** Time delays and **b** lost deadlines, with light traffic (50–80 s)

The behaviour of the network with light traffic generates an increment in the behaviour of the time delay, as well as lost deadlines. The time delay has a maximum of 27 ms, with an average of 10 ms, and a 4.07% of lost deadlines with an average of 146 ms between lost deadlines. Figure 6.33a shows the time delay, in which the light traffic starts at 50 s. However, the effect initiates a few seconds later, due to packet flooding of the network. On the other hand, the light traffic finishes at 80 s, but its effect remains up to 90 s, due to the switch continues taking packets from its queue. Something similar happens with the lost deadlines, in which the packets lose deadlines due to queueing.

As in the previous case, the fuzzy scheduler evaluates the system's and network's performance at each scheduling period $\delta = 1$ s and defines the next sampling period h. When light traffic is present, the MAE (Fig. 6.34a) presents light variations that modify the fuzzy scheduler behaviour, with a 4% of lost deadlines (Fig. 6.34b). Thus, the behaviour of the fuzzy scheduler with light traffic increases the sampling period with a smaller rate when there is a small error, aiming not to degrade system's performance due to the lack of information, because of the presence of lost deadlines.

In the next case, system's behaviour is observed when there is a medium traffic in the rank of [50 80] s. Although both approaches remain stable, in this case, there is a clear difference between the performance of the fuzzy codesign and the performance of the hybrid control. Fuzzy codesign maintains the control criteria with medium traffic (Fig. 6.35a), with an overshoot of $\zeta = 15°$, an average error at stable state $e_{ss} \approx 0.5$, and a steady time $T_s = 4$ s. On the other hand, hybrid control (Fig. 6.35b) degrades its performance with medium traffic, presenting oscillations, and exhibiting an overshot of $\zeta = 34°$, an average error at stable state $e_{ss} \approx 6$, with oscillations. At this level of traffic, hybrid control assures system stability, but with a bad performance (Fig. 6.36).

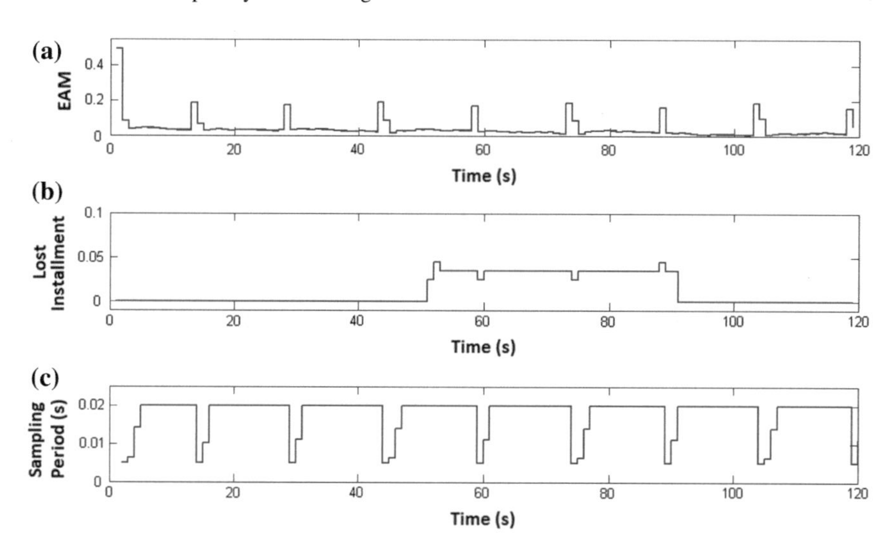

Fig. 6.34 NCS **a** MAE, **b** lost deadlines, and **c** sampling period, with light traffic

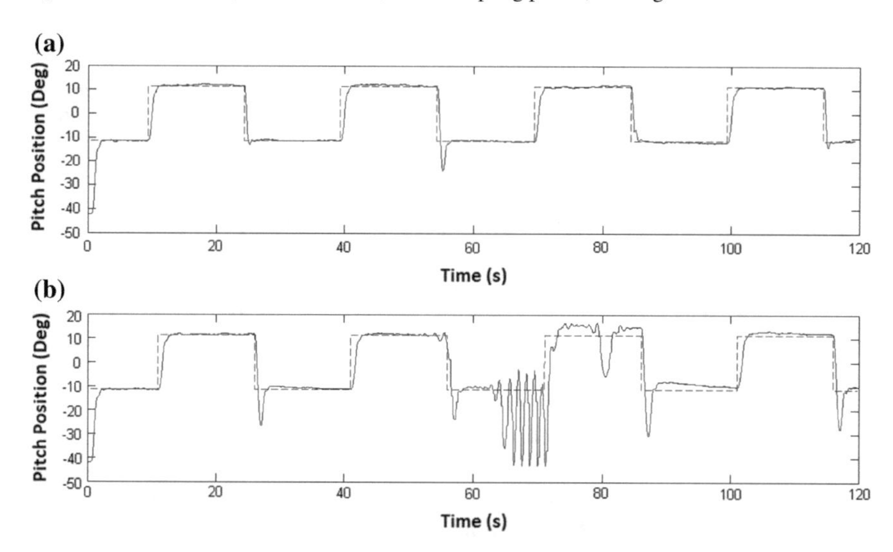

Fig. 6.35 Performance of **a** the fuzzy codesign and **b** hybrid control, with medium traffic (50–80 s)

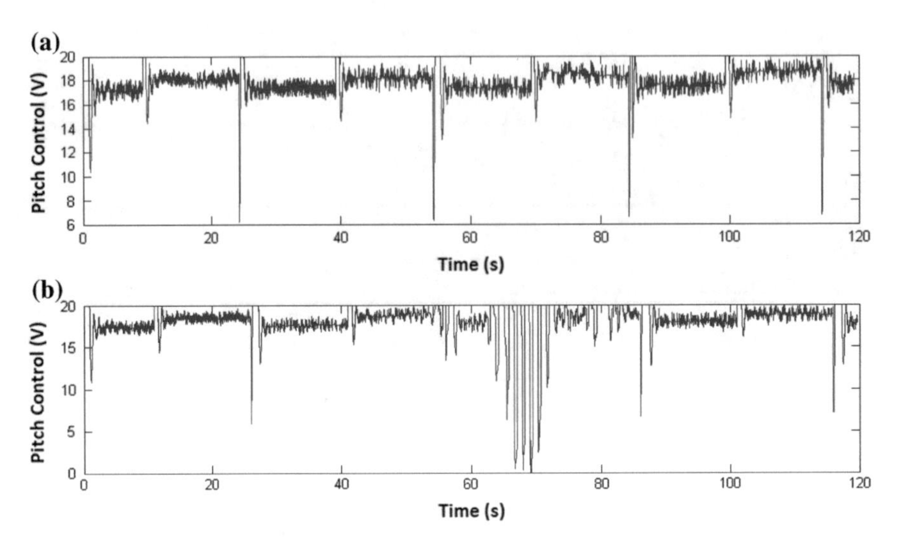

Fig. 6.36 Control signal of **a** the fuzzy codesign and **b** the hybrid control, with medium traffic

The bad performance of the hybrid control is due to abrupt changes in an ample rank of operation since the traffic produces a faster dynamic in the controller. The fuzzy codesign, in contrast, switches with a slow dynamic, performing small changes in the control signal.

The network behaviour with medium traffic generates a peculiar behaviour in the time delay and the lost deadlines. The time delay increases at the beginning of the traffic application with a maximum of 31 ms, an average of 5 ms. However, while traffic increases, time delays become lost deadlines with 5.8%, and an average time of 132 ms between lost deadlines. Figure 6.37a shows the time delay behaviour, in which traffic starts at 50 s. However, the effect initiates a few seconds later, and decreases at 55 s, increasing the lost deadlines at 80 s. Nevertheless, its effect remains until 85 s, due to the switch keeps on taking packets from its queue.

In this case, the fuzzy scheduler changes the policy for assigning the sampling period (Fig. 6.38c). In the presence of medium traffic, MAE presents variations larger than 0.1 (Fig. 6.38a), that represents an error at stable state greater than 2%, along with the presence of 5.8% of lost deadlines (Fig. 6.38b). Here, the fuzzy scheduler slowly decreases the sampling period, but without overloading the network, thus gradually improving system's performance.

In the case of heavy traffic, the hybrid controller presents an erratic behaviour, while the fuzzy codesign performance keeps stable, maintaining the control criteria. Fuzzy codesign with heavy traffic keeps an overshot of $\zeta = 30°$, but during a very short time (3 s), with an average error at stable state of $e_{ss} \approx 0.8$, and a steady time $T_s = 4$ s (Fig. 6.39a). In contrast, the hybrid control degrades its performance in the presence of heavy traffic, with an overshoot of $\zeta = 36°$, also, during a very short time (3 s), with an average error at the stable state of $e_{ss} \approx 10$, with oscillations (Fig. 6.39b). The bad performance remains, even when the heavy traffic is gone.

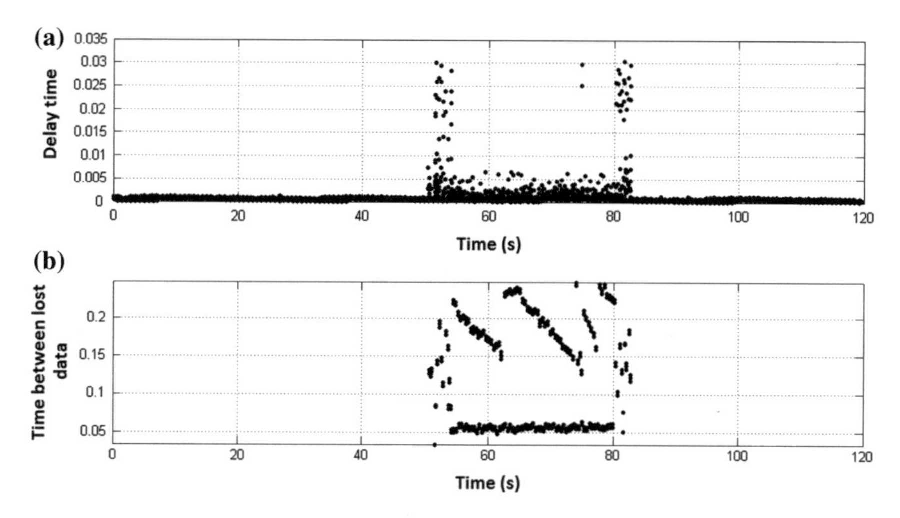

Fig. 6.37 **a** Time delays and **b** lost deadlines, with medium traffic (50–80 s)

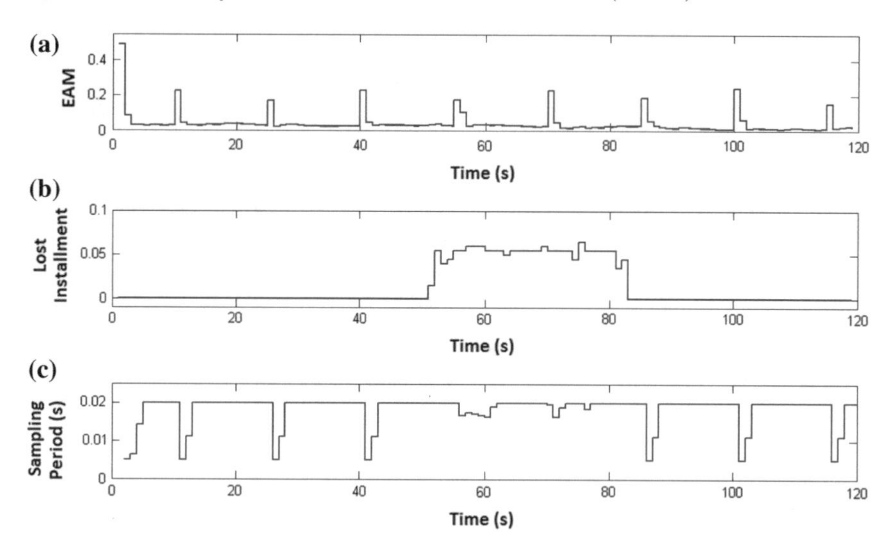

Fig. 6.38 NCS **a** MAE, **b** lost deadlines, and **c** sampling period, with medium traffic

With this case, it is shown that the fuzzy controller is more robust against greater variations in the network uncertainties than the hybrid controller.

The control signal behaviour for both approaches is very similar in the case of heavy traffic. The hybrid controller performs abrupt changes due to the fast dynamic of the controller (Fig. 6.40b), while the fuzzy codesign switches with a slower dynamic, carrying out small changes in the control signal.

In the presence of heavy traffic, both time delay and lost deadlines increase. The maximum time delay is 30 ms, with an average of 12 ms, and a 7.3% of lost deadlines,

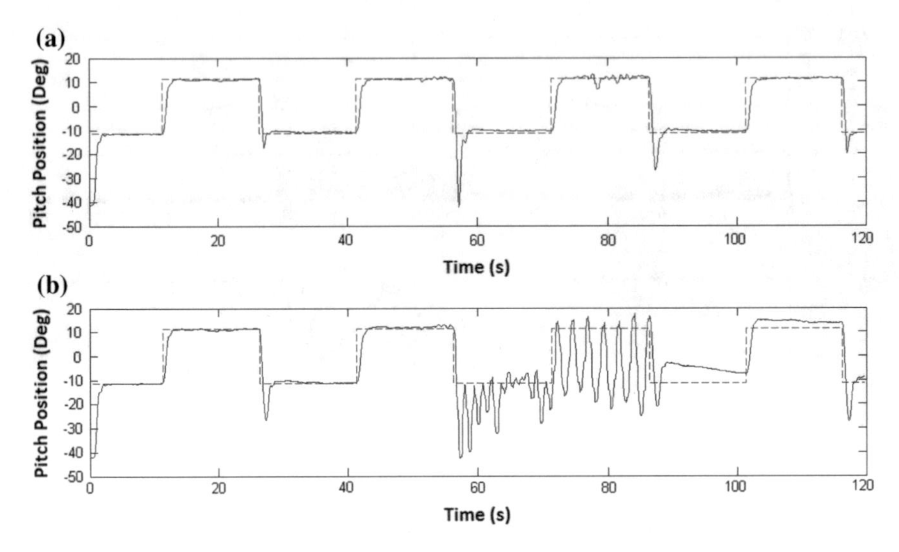

Fig. 6.39 Performance of **a** the fuzzy codesign and **b** hybrid control, with heavy traffic (50–80 s)

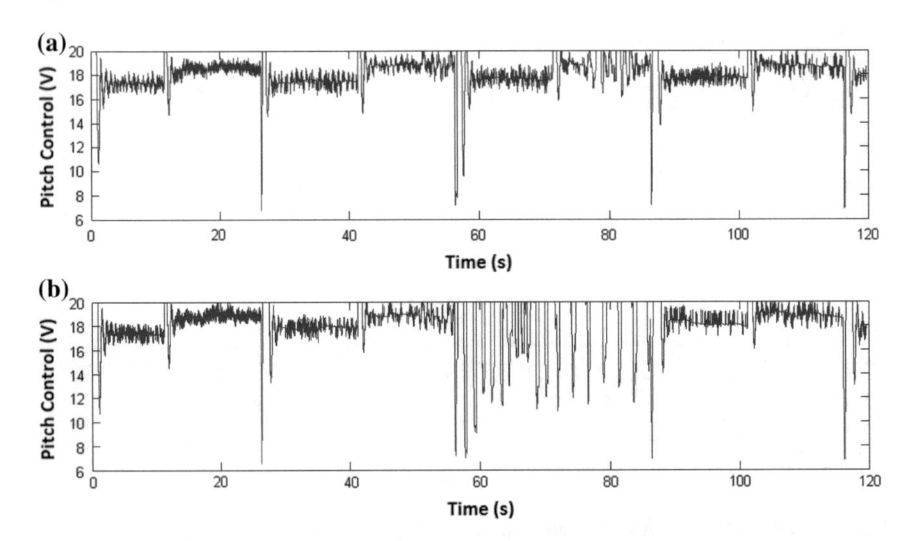

Fig. 6.40 Control signal of **a** the fuzzy codesign and **b** the hybrid control, with heavy traffic

with an average time between lost deadlines of 122 ms. Figure 6.41a shows the time delay, in which heavy traffic starts at 50 s, and its effect remains until 85 s, even though the traffic finishes at 80 s.

With heavy traffic, the fuzzy scheduler changes the sampling period assignation policy (Fig. 6.42c). This due to the MAE presents variations greater than 0.25, which represent an error at stable state greater than 5%, along with the presence of 7.8%

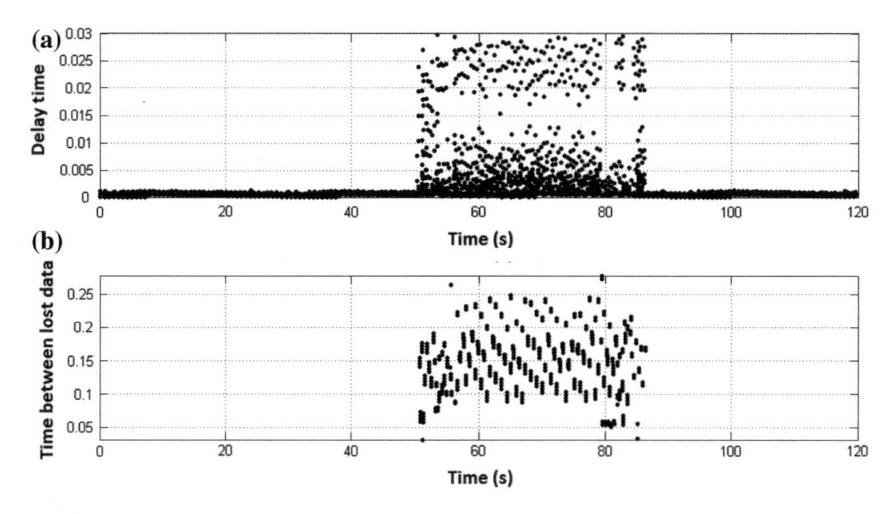

Fig. 6.41 **a** Time delays and **b** lost deadlines, with heavy traffic (50–80 s)

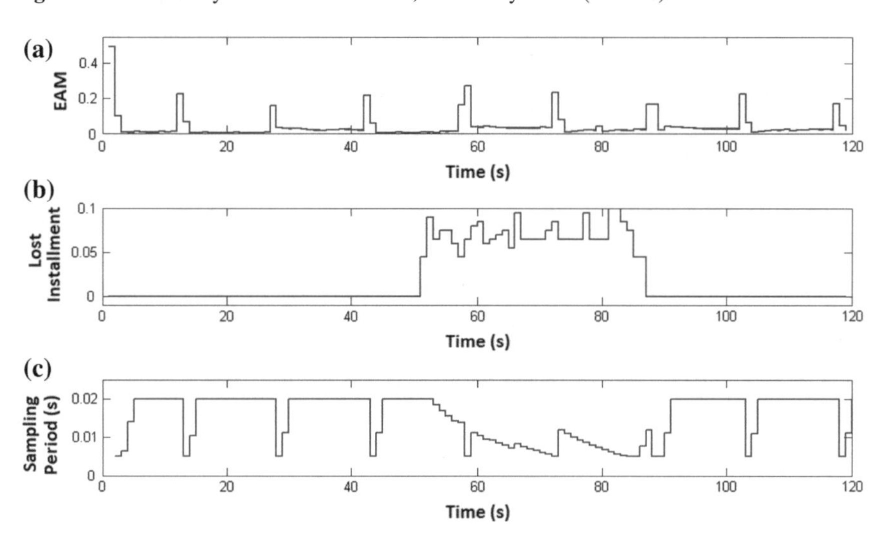

Fig. 6.42 NCS **a** MAE, **b** lost deadlines, and **c** sampling period, with heavy traffic

of lost deadlines (Fig. 6.42b). The fuzzy scheduler slowly decreases the sampling period, to correct the error even when network traffic increases.

6.5 Concluding Remarks

As a result of this control design, it is possible to realise how the effects of time delays modify the behaviour of the dynamic system in several scales, and thus, in several scenarios, such as bounded time delays and lost packets. It is important to guarantee the stability of the dynamic system during several time delays stages, like for example, undersampling, among others. Several results presented in this chapter are related to those bounded time delays modelled through the techniques viewed in Chaps. 3 and 5, therefore, presenting a challenging response to these planned situations.

In the SISO MAGLEV example, the generalised exponential distribution to estimate variable sampling periods is proved with real variable sampling periods, generated in the Ethernet network. Results show an adequate estimation for use in the statistical model.

The design of a fuzzy controller is proposed, aiming to stabilise a nonlinear NCS. This design is optimal for a static sampling period, and asymptotically stable in the large with variable sampling periods longer than a base sampling period. This control is robust to several lost data and larger time delays into the network.

The design of a sampling frequency scheduler is proposed, this modifies the sampling frequency of each sensor node to adjust the transmission frequencies according to the QsS performance. The example shows a better scheduling of sampling tasks with a reduction of lost packets and time delays.

The best practice is a codesign where simultaneously a control and scheduling design is performed. In this chapter both designs considering time delays larger than sampling period and lost packets, controlling the actuation period and signal control and getting a better QsC and QsS performance than a hybrid methodology.

Future work needs to be carried out, in order to understand the communication network as a dynamic system. Furthermore, a better understanding is necessary when intermittent long delays appear into the network, as well as get bounds for the time delays and packet loss.

References

1. Mendez-Monroy, P.E., Benitez-Perez, H.: Fuzzy control with estimated variable sampling period for non-linear networked control systems: 2-DOF helicopter as case study. Trans. Inst. Measur. Control **34**(7), 802–814 (2012)
2. Heemels, M., Van de Wouw, A.N., Nesic, D.: Networked control systems with communication constraints tradeoffs between transmission intervals, delays and performance. IEEE Trans. Autom. Control **55**(8) (2010)

Chapter 7
Conclusions

This book has presented several strategies for time delays modelling through the use of real-time scheduling algorithms or based period measurements and frequency transmissions from communication channels. Further, a review of control design based upon fuzzy Takagi-Sugeno-Kang (TSK) approximation is presented including time delays behaviour.

Current time delays can be modelled using real-time dynamic scheduling algorithms; however the resulting delays are time varying and stationary, therefore related local control laws need to be designed according to this characteristic and time integration is the key global issue to be taken into consideration. Global stability is reached by the use of TSK Fuzzy Control Design where a nonlinear combination is followed by the current situation of the states which are partially delayed due to communication behaviour. Therefore Fuzzy Control may be attractive to guarantee global stability since any condition is bounded to be less than sampling period at the worst case scenario with no loose of generality. The use of dynamic scheduling approximation allows the system to be predictable and bounded; therefore, time delays can be modelled in these terms.

Moreover, the resulting dynamic representation tackles the inherent switching per scenario. This approximation has the main drawback that context switch may be invoked every time a periodic task takes place and it is possible to be executed; in this case, inherent time delays to this action are taken into account to be processed as uncertainties.

In a NCS, time delays can be modelled using real-time dynamic scheduling algorithms; however the resulting delays are time varying and stationary, therefore, related local control laws need to be designed according to this characteristic. Time integration is the key global issue to be taken into consideration as well as local variable structure.

Global stability is reached by the use of TSK fuzzy control design, where a nonlinear combination is followed by a current sequence of states, which are partially

H. Benítez-Pérez et al., *Control Strategies and Co-Design of Networked Control Systems*, Modeling and Optimization in Science and Technologies 13,
https://doi.org/10.1007/978-3-319-97044-8_7

delayed due to the communication behaviour. Since this powerful technique such as Fuzzy TSK, it is possible to guarantee but features time delay response as well as local variable structure. Priority exchange allows a system to be known in terms of local time delays. The use of a dynamic scheduling approach allows the system to be predictable and bounded, and therefore, time delays can be modelled in these terms.

Moreover, the resulting dynamic representation tackles the inherent switching per scenario. However, this approach has the main drawback that context switch may be invoked every time that an aperiodic task takes place, and it is possible to be executed. In this case, inherent time delays to this action are retaken into account to be processed as uncertainties.

A method is presented to estimate the variable sampling period of a NCS that is the sum of sensor-controller time delay and packet losses with a controller-actuator estimated the sampling period with a probability density function. The model is computationally simple and light, as necessary for storage of data of sensor-controller time delays. The estimation of variable sampling period is applied within an Ethernet network with multiple traffic nodes showing with experimental data a good estimation useful for control.

Furthermore, the design of a fuzzy model from a nonlinear model to compensate for the variability of sampling period is presented, where this model selects the best discrete model to estimate the next system state as a function of the variable sampling period. Using this fuzzy model, a fuzzy controller is designed to stabilise the NCS; this design is optimal for a static sampling period and globally stable with variable sampling period longer than a base sampling period but bounded. A non-linear, unstable MIMO system is used for the case study, where a controller was designed using the new method to prove stability and robustness to time delays and packet loss. Three tests are presented.

The first test shows the stability of a control loop without traffic into a network but with a wide range of variable sampling periods with a similar behaviour to digital control without a communication network. The second test shows stability and robustness to traffic incorporating four traffic nodes, generating a larger range of sampling periods and packet losses, obtaining a better performance than feedback control and a similar performance to the first test. Finally, the third test shows robustness to a considerable percentage of packet losses and a wide range of variable sampling periods, obtaining a better stability and better performance than feedback control that has persistent oscillations and bad performance.

This work deals with the available transitions for achieving schedulability. Saturated conditions are not considered. We have presented a linear time-invariant model of nodes and their frequency transmission as involved in a distributed system. The significance of controlling the frequencies stems from the system schedulability. The key feature of the LQR control approach is a simple design with good robustness and performance capabilities that allow the frequencies to be easily modified. We have shown via numerical simulations the performance of the proposed control scheme. In both cases, the case study and the network controller, it is possible to show the actual feasibility of this co-design strategy since a complex problem of non-coupled dynamics is presented.

Further study is needed in order to propose a holistic integration of both dynamics that can be developed through high order observers. This work would be of special interest. Alternatively, the Coordinated Task, as a protocol to perform a distributed task on MANET networks. The distributed task is executed as a response to events in the network. An event is any variable sensed by any node of the network.

Particularly, the protocol discussed here is suitable when a temporal new behaviour is required due to a specific event in the environment. The simulation results show that the density of nodes has a higher impact on the execution time. For a larger density of nodes, there is longer execution time. Moreover, the latency of the execution time also increases with the density of nodes. According to this, it is suggested that the effects of the density of nodes require further study. How many nodes should execute the same process, and how many nodes are required in total to guarantee the execution, are left as future work. Another issue regarding the use of this protocol is how to deal with missing nodes, due to the nodes leaving the network. For this, the protocol proposes looking at the local cache for alternative nodes. However, the results show a low rate of successfully executed Coordinated Tasks, while the number of processes increases. An alternative to explore is to allow the nodes to search in the cache of their neighbours.

Glossary

Abstraction A representation or description of a system based on the assumed essentials, filtering those features which do not seem relevant.

Address space Collection of addresses in memory in which an object or process exists during its execution.

Aggregation The activity of adding independent components to another component, hence creating a larger, composed component. The gathering together related elements, with close functions or purposes.

Application framework Set of components assembled together, which interact in order to represent a software structure (or coordination) of related applications.

Asynchronous I/O Mechanism for sending and receiving data using I/O operations, in which the sender does not block waiting to complete the sending operation.

Bandwidth The amount or capacity of a particular communication media, such as a network, bus, or any other media.

Broadcast A method for disseminating a particular information from a sender to several receivers of a distributed environment.

Bus Set or array of physical communication links between devices, such as processors, memory, I/O devices, etc.

Busy waiting A state of a process in which such a process is executing a loop waiting for a condition to be true.

Cluster Of Workstations (COW) A cluster is composed by nodes, and each node is a complete workstation except for the common I/O peripherals (monitor, keyboard, mouse, etc.). However, nodes normally have a local disk. A node may be a simple Personal Computer, along with its processor, local memory, and network interface, or it could even be a full SMP computer. Nodes communicate with each other through a low cost commodity network like Ethernet, ATM, or others.

H. Benítez-Pérez et al., *Control Strategies and Co-Design of Networked Control Systems*, Modeling and Optimization in Science and Technologies 13,
https://doi.org/10.1007/978-3-319-97044-8

Communications In order to cooperate, parallel tasks require exchanging data. There are several ways to accomplish this, such as through shared memory or over network. However, the actual event of data exchange is commonly referred as communications, regardless the method employed.

Complexity A measure of the number of types of internal relationships among the components of a system. The more complex a system is, the more difficult it is to design and build.

Concurrency A characteristic of a process, an object, a component, or a system to execute operations simultaneous, this is, with the potentiality of simultaneous execution, but just emulating such an execution.

Condition synchronization Synchronization technique that involves delaying a process until its state satisfies a particular condition, normally a Boolean condition.

CPU Central Processor Unit, a set of digital circuits that perform the operations on data as described by instructions. Both data and instructions are stored in memory.

Device Hardware component that provides services for computing or communication.

Distributed computing Programming activities related with the design and implementation of an application that allocates processes, objects, or sub-systems on the nodes or computers of a network.

Distributed memory Network based memory access for physical memory that is not common. As a programming model, tasks can only logically use local computer memory, and must use communications (I/O operations) to access memory of other computers, where other tasks are executing.

Distributed program The specification of software components that communicate using message passing, remote procedure calls, or rendezvous. Usually, the software components execute on different, distributed computers within a network.

Distributed Shared Memory (DSM) Implementation of a shared memory space executing on a distributed memory multiprocessor or network of computers.

Earliest deadline first A single resource scheduler that allows ordering tasks activity based upon priorities following time response and the earliest deadlines amongst tasks.

Event Message that contains the occurrence of a significant action, along with the data relevant for such an action.

Heterogeneous Systems based on different components with specialized behavioral rules and relations. Basically, the operation of the system relies on the differences between components, and therefore, no component can be swapped with another. In general, heterogeneous systems are composed of fewer components than homogeneous systems, and communicate using function calls.

Homogeneous Systems based on identical components interacting in accordance with simple sets of behavioral rules. They represent instances with the same behavior. Individually, any component can be swapped with another without noticeable change in the operation of the system. Usually, homogeneous systems have a large number of components, which communicate using data exchange operations.

Inter-Process Communication (IPC) Mechanism that allows the communication between processes that reside in different address spaces.

Kernel Set of data structures and primitive atomic operations that manages and controls processes, scheduling them on processors, and implementing high-level communication and synchronization operations, like semaphores or message passing.

Maximum allowable transfer interval This is the longest time within a node should transmit a data within a periodic interval.

Message Fundamental unit of communication between threads, processes, objects, components, sub-systems, and systems.

Message passing An inter-process communication and synchronization mechanism between two or more parallel or distributed software components, executing simultaneously, non-deterministically, and at different relative speeds, on different address spaces of different computers of a distributed memory parallel platform.

Model An abstract representation of some aspect or aspects of a system.

Modeling Creation of a model as an abstract representation of actual systems.

Network Communication hardware that allows the connection among computers.

Networked control systems Networked control systems refer to a dynamic system that uses a communication network to cooperate amongst sensors, controller and actuators in order to produce a corrective signal within dynamic system.

Network interface Hardware device that connects a network to a computer.

Node A computer system that represent the basic unit or component within a network or distributed system.

Operating system Collection of services and APIs that manages hardware and software resources of a computer system.

Parameter Instance of a data type which is passed to a function or method.

Process Interaction between the processor and memory, in which the processor is instructed what operations to perform on what data by a sequence of instructions.

Rate monotonic Real time single resource scheduler that allows tasks ordering based upon priorities following a basic relation between local consumption and periodic tasks.

Real tme Real time is a very productive concept based upon the idea of inherent processing time of local processes achiving deadline in a specific time measurement according to inherent clock.

Real-time distributed systems this type of systems belong to a specific time response within local and global entities. The restrictions may be quite varied but are common to time response following several craitiries like global clock, tide sincronization, time stamping or local time aggrements.

Request event A message sent by a client to a server asking for a service.

Round trip time this is the time elapsed between two nodes consumed during communication procedure and it is the add of several local time delays.

Scheduler Mechanism which decides the order in which threads, messages or events are executed by a processor.

Scheduling policy Policy that determines which action is the next to be executed, this is, establishes the order in which operations execute.

Server Application or computer host that provides services.

Service Functionality offered by a server to a set of clients.

Shared memory A computer architecture where all processors have direct (usually bus based) access to common physical memory. In a programming sense, it describes a model where parallel tasks all have the same blocks of memory, and can directly address and access the same logical memory locations regardless of where the physical memory actually exists.

Sub-system A set of cooperating components that cannot be considered a whole or complete system, but performs a defined functionality independent from the rest of the system.

Synchronization Coordination of parallel tasks in real-time, often associated with communications among them. Often implemented by establishing a synchronization point or mechanism within an application where a task may not proceed further until another task(s) reaches the same or logically equivalent point. Usually, it involves at least one task waiting, which means that the wall-clock execution time of a parallel application to increase.

Synchronous I/O Mechanism for sending and receiving data using I/O operations, in which the sender does block waiting to complete the sending operation.

System The collection of components, their connections (or relations), and their organization rule or form, which as a whole, perform a function that cannot be achieved by the individual components.

Task A logically discrete piece of computational work. A task is a set of instructions executed by a processor.

Thread A sequential set of instructions or operations that are performed by a single control or context, and thus, is able to execute concurrently with other threads. Threads can be executed on a single processor, competing for time, or can be executed in parallel on separate processors.

Printed in the United States
By Bookmasters